LQ7 32.86 + 10 46 ex
W9-CKB-807

SOIL–PLANT RELATIONSHIPS
An Ecological Approach

SOIL~PLANT RELATIONSHIPS

An Ecological Approach

DAVID W. JEFFREY

CROOM HELM
London & Sydney

TIMBER PRESS
Portland, Oregon

© 1987 David W. Jeffrey
Croom Helm Ltd, Provident House, Burrell Row,
Beckenham, Kent BR3 1AT
Croom Helm Australia, 44–50 Waterloo Road,
North Ryde, 2113, New South Wales

British Library Cataloguing in Publication Data
Jeffrey, David W.
Soil–plant relationships: an ecological approach.
1. Botany — Ecology
I. Title
581.5′26404 QK901
ISBN 0–7099–1459–8
ISBN 0–7099–1464–6 Pbk

First published in the USA 1987 by
Timber Press,
9999 S.W. Wilshire,
Portland, OR 97225,
USA

All rights reserved

ISBN 0-88192-075-0 (cloth)
ISBN 0-88192-076-2 (paper)

Printed and bound in Great Britain
by Billings & Sons Limited, Worcester.

Contents

Preface

Soil–plant relationships once had a limited meaning. To the student of agriculture it meant creating optimum conditions for plant growth. To the ecologist it meant explaining some plant community distribution patterns by correlation with soil type or conditions. This dual view has been greatly expanded at an academic level by the discovery of the ecosystem as a practical working unit. A flood of concepts and information subsequently emerged from the International Biological Programme. At a totally different level of resolution, it is appreciated that certain soil-based ecological problems have a molecular basis, and must be addressed by physiological or biochemical approaches. From ecosystem to molecule we have powerful new tools to increase the flow of ecological data and process it for interpretation.

Society is now experiencing a series of adverse global phenomena which demand an appreciation of soil–plant relationships. These include desertification leading to famine, soil degradation accompanying forest destruction, acidification of watersheds and the spasmodic dispersal of radionuclides and other pollutants. It is public policy, not merely to identify problems, but to seek strategies for minimising their ill effects.

This book is written as a guide to soil–plant relationships, centrally oriented towards ecology, but of interest to students of geography and agriculture. For ecology students it will bring together subfields such as microbiology, plant physiology, systematics and provide interfaces with animal biology, meteorology and soil science. Ideas contributing to the formulation of hypotheses are emphasised, rather than overloading readers with information better obtained first hand from the primary literature or encyclopaedic reviews.

The selection of case histories in ecological investigation indicates the range of possible approaches to demonstrate that ecology is a positive and applicable science, capable of earning its keep in the eyes of the world.

I must gratefully acknowledge the contributions made to my ideas by past and present colleagues in Ireland, Britain, Australia and North America. They are too numerous to name individually and their generosity has been unbounded. However, I owe most in terms of concept development and stimulation to my students and my family.

David W Jeffrey
Dublin 1986

Part One

A plant-centred biological complex

The theme of this section is the response of plants in relation to ions and water in the soil environment. It draws special attention to the close biological relationships between plants, microorganisms and other organisms in the food web. The role of plants as primary producers is taken for granted, even though limits to plant growth imposed by soil conditions will be encountered many times. This account should be read in conjunction with the prolific literature on production ecology.

1

Plants, roots and ion absorption

INTRODUCTION

The view of ecology adopted here is that it comprises the study of ecosystems and their surroundings. Plants are thus seen in relation to the cluster of organisms most closely associated with them and to their environment. Plants are located in an environmental complex of energy, atmosphere and soil. Within an ecosystem context, plants have three principal functions:

(a) Absorption of photosynthetically active radiation which is applied to the synthesis of carbon-to-carbon bonds. Compounds synthesised in this way serve as the only energy currency for the ecosystem.
(b) Absorption and assimilation of ions as a source of essential elements for the ecosystem. Photosynthesis is necessary for ion absorption and assimilation, and ions are an essential part of the photosynthetic system and all other metabolism.
(c) Plants process water on a sufficiently large scale to make a large contribution to the hydrological cycle of the Earth. Some of this water use is ancillary to photosynthesis, general metabolism and the fabric of plant structure. Much is involved with the special place of plants in unavoidably intercepting solar heat while simultaneously optimising gas exchange. Leaves are cooled by evaporative water loss.

These activities are represented in a simple way by Figure 1.1, which symbolises leaves and roots configured for their respective functions. Storage of energy and ions is also regarded as a universal feature.

Figure 1.1: Plants are organisms that are rooted in soil and receive and process energy, water and ions. They are the foundation of ecosystems and contribute to hydrologic and geochemical processes

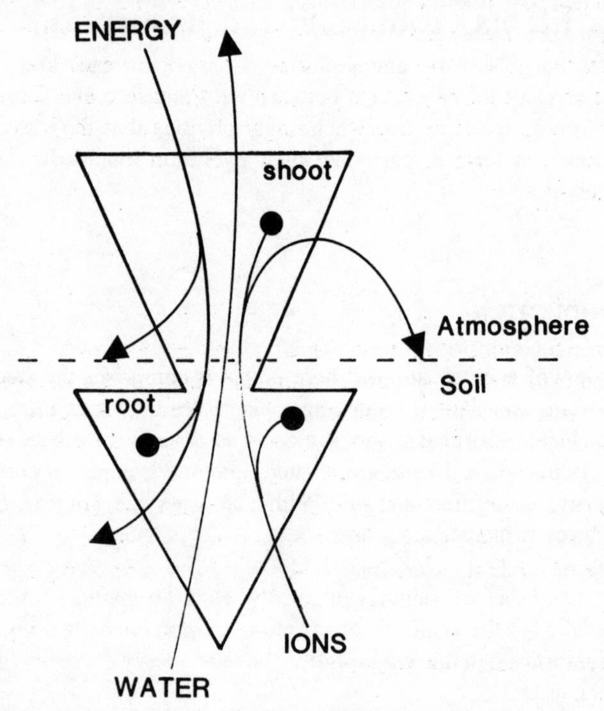

The ecologist should observe the vegetation being studied with the eye of the designer, the systems analyst, the civil engineer and even the artist. This appraisal will prove valuable in hypothesis building. The simple realisation formalised by Raunkier, that plants' life forms can be readily classified and that whole communities have common morphologies and growth behaviours, is still important. It makes us realise for a start that sheer maximisation of biomass production is not a characteristic of predominant ecological importance. Long-term persistence of a species is much more subtle, entailing the optimisation of: assimilating the materials for primary production; producing and disseminating propagules; accommodating to the short-term variations in environment; accepting predation; and adapting to long-term environmental fluctuation and even cataclysm.

The means for long-term persistence under particular soil-environment regimes entails a range of features including morphological and physiological plasticity, ecotype formation, species characteristics and indeed the collective properties of larger taxa. Ecologists will broadly recognise the association between the vetches and peas (Papilionoideae) and nitrogen fixation; between the heathers (Ericaceae) and survival on infertile, acid substrates; between the glassworts and seablites (Chenopodiaceae) and succulence and salt tolerance; and between the Cactaceae and survival in arid zones. It seems overwhelmingly obvious that the vascular plants have, at least in part, had their evolution moulded by soil characteristics.

ROOTS

The three functions of all roots which are outwardly directed to the environment, and which are worth discussing in an ecological context, are absorption, anchorage and storage. Storage may, according to some opinions, fall into the category of internal metabolic functions for roots, which include, for example, the production of growth regulators such as kinetin and abscisic acid (Russell 1977). However, it is not possible to discuss the absorption and fate of mineral metabolites without reference to storage.

Root growth and root structure

Unlike the complex apex of the shoot, the root apex is small and compact. It cuts off files of cells which differentiate to produce epidermis, cortex and vascular bundles behind the apex and cells of the root cap before it. The first stage of cell differentiation is elongation. The energy for this process is generated from the translocation of osmotically active metabolites, sugars or ions, to the vacuoles of the expanding cells. Entry of water and development of a substantial pressure potential elongates the cells. Orientation of cellulose fibrils in cell walls, which resembles that of the hoops of a barrel, coordinates the direction of growth. This uniaxial extension growth can be thought of as a jacking action, with a force of up to 10 bars (1.0 MPa), which drives the root cap through the soil, reacting against the frictional and other forces gripping the older parts of the root.

The dimensions of a typical root apex, in the order of 0.2 mm diameter, are sufficiently small to penetrate most inter-aggregate spaces in soil (Figure 1.2). The root cap cells appear to be physically sloughed off by the passage of the apex through soil. These cells can produce a copious secretion of 'mucigel' (well studied in cereals), which may well have a range of functions:

(a) lubrication for the movement of the apex through soil:
(b) establishment of liquid film contact between root and soil aggregates;
(c) encouragement of a rhizosphere flora of bacteria and fungi.

The root cap itself appears to be responsible for geotactic growth of roots.

Maturation of the root proceeds by the differentiation of the stele, an endodermis and the pericycle enclosing xylem and phloem strands. At the point when the maturation process is complete, root hairs appear on the root surface. These are, relatively speaking, enormous extensions of the cell wall of the epidermal cells. The dimensions resemble those of fungal hyphae in general order of magnitude of diameter, i.e. 10–15 μm. A frequently quoted length is 1000 μm (1 mm). (A 1 metre length of garden hose has similar diameter-to-length proportions to root hairs.) Persistence of the root hairs in seedlings is usually limited to a few days. This means that the root hair zone, in an actively extending root, is being propelled through soil in a manner resembling a bottle brush being pushed into a bottle. Longer root hair life, and extensive root hair cover are common in many perennial plants. Make a point of examining a small portion of freshly collected roots of many common species with a × 10 magnification lens to verify this. Although root hairs can absorb water and ions, agricultural research, e.g. Russell (1977), emphasises that their presence is not essential for either process. The performance of root hairs under non-agricultural conditions, i.e. in conditions of low fertility, is very unclear. Root hair proliferation under conditions of low fertility is recognised occasionally (see Chapter 16). The root hair is frequently the point of infection by root nodule bacteria, vesicular–arbuscular mycorrhizal fungi or root pathogens. From the ecological point of view it is important to realise that the development of the root microflora is parallel and complementary to the growth of the plant structure.

Lateral roots develop as initials within the stele which then emerge through the cortex virtually at right angles to the main root axis. Production of secondary laterals may proceed in the same way. We take for granted that initiation of laterals is under metabolic

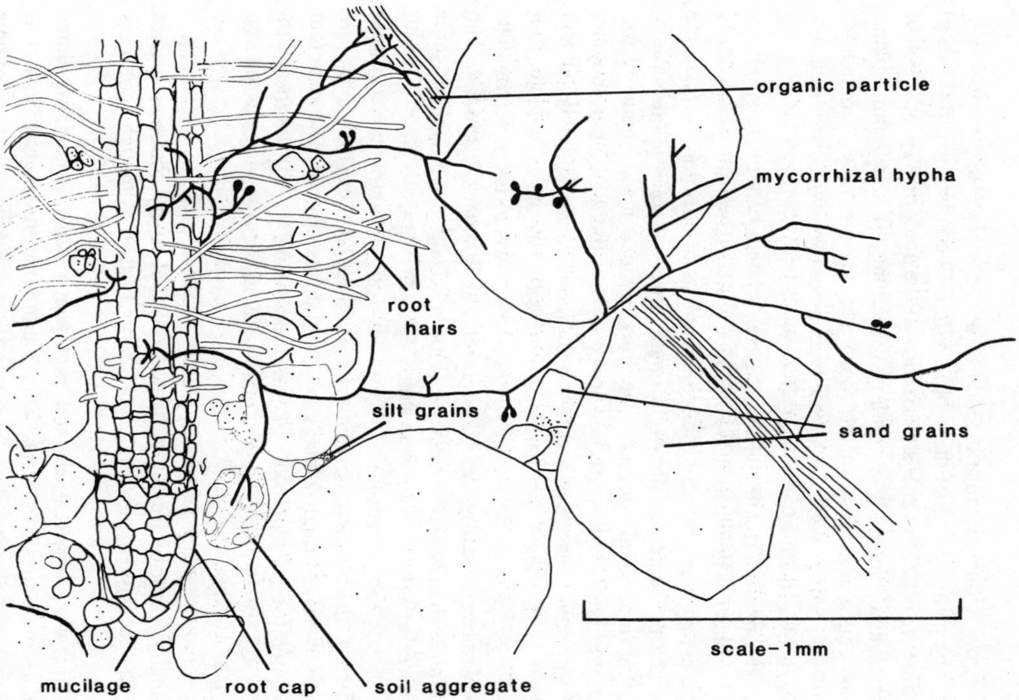

Figure 1.2: Root structure in relation to soil-aggregate structure. This sketch depicts the root tip of a herbaceous plant during the growing season. It is depicted with root hairs and a vesicular–arbuscular mycorrhiza. The soil is simplified to give a sense of scale, and other soil organisms are omitted. (Information from many sources.)

control, if only because it tends to be quite regular (Figure 1.3).

The primary axis of the root in a long-lived plant will proceed to mature further, towards suberisation of the epidermis, production of fibres and eventually secondary thickening. In a perennial plant, senescence and decay of the root system is probably a continuous process, with peaks of senescence associated with any seasonal arrest of growth of the plant as a whole.

The overall architecture of the root system of a given species is very plastic (Feldman 1984). Roots will respond morphologically and metabolically to soil conditions such as compaction, wetness and aeration. Compact soils will ultimately prevent penetration of roots, acting in effect as bedrock. The more compact a soil, the more frequently lateral roots are developed. Thus, the upper horizons are more actively explored and root penetration causes soil development from the surface. The distribution of water in soils, for example a falling capillary fringe, may lead roots down to great depths. The slow movement of water in unsaturated soils means that root growth is faster than water movement. Root penetration to a depth of some metres is known in dune systems and grasslands.

When low aeration conditions develop, root structure can alter towards the formation of aerenchyma. Root cortex cells lyse to form open canals through which gases may diffuse freely. The source of oxygen for growing roots may well be the interior of the root, with some 12% of the cross-sectional area as air spaces (Jackson and Drew 1984).

ROOTS AND SOIL

Plants are exposed to the soil environment through their roots and other underground organs. This complex interface must be seen as fully three-dimensional. A vegetation unit is dependent on a volume of soil for supply of water and mineral resources and for physical support. One should become aware of the extent of roots of species and the general rooting depth of vegetation. This is relatively easily perceived in non-woody vegetation and from observation of wind-felled forest trees. Wind felling usually happens when a strong wind coincides with water-soaked soil with low strength. Structural failure then occurs at the edge of the root mass rather than through it. Any field observations of roots will also immediately show that root distribution is not uniform through soil, root biomass being greatest close to the surface. At a very conservative estimate it is

Figure 1.3: Root morphology of idealised young herbaceous plant

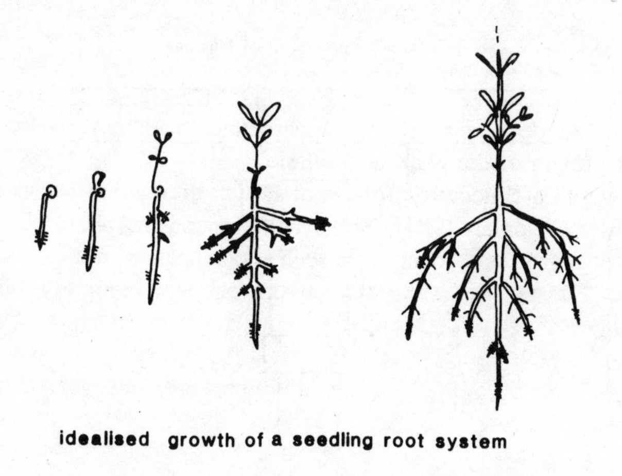

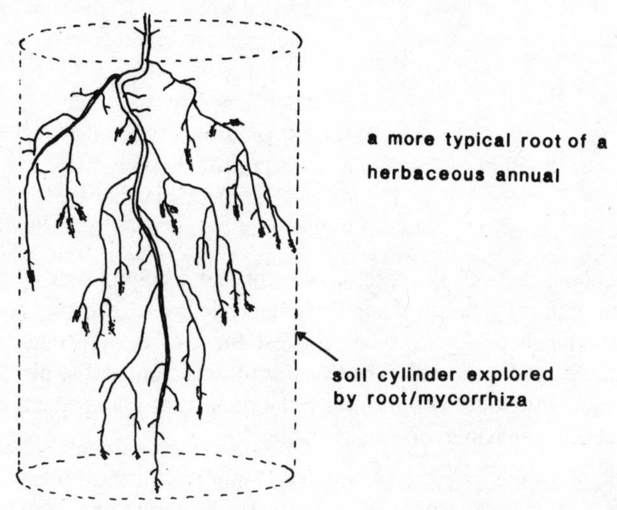

probably true that 75% of roots are found in the upper half of the rooting zone. The data set from northern shortgrass prairie (Coupland 1979) illustrates this point for a particularly deep-rooted vegetation type (Figure 1.4).

The exploration of soil by roots matches the ordered array of leaves and shoots in the aerial environment. The plant is thus bridging a continuum between soil and atmosphere. The soil–plant–air

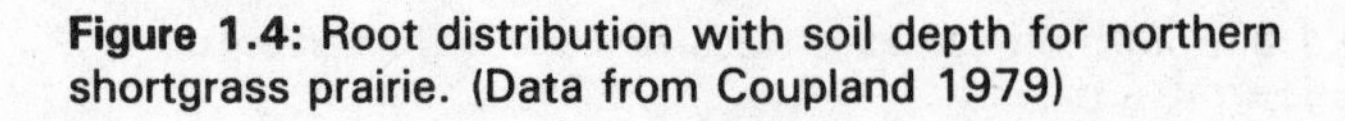

Figure 1.4: Root distribution with soil depth for northern shortgrass prairie. (Data from Coupland 1979)

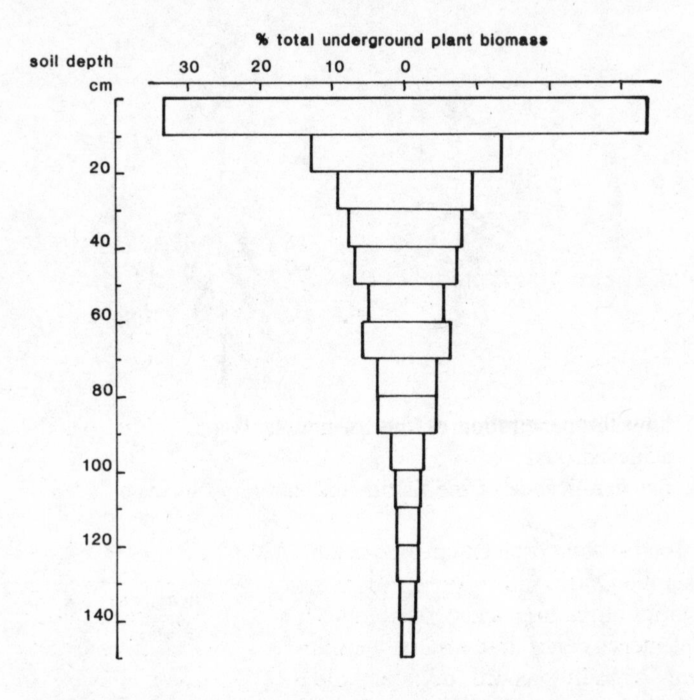

continuum (SPAC) is a useful concept and a helpful unit of study when exploring the exchange of water, ions and energy between environment and plant. The simplest form of these exchanges is illustrated in Figure 1.5. The transfer of ions from soil to plant will be dealt with under two headings, the process of ion uptake and the metabolic behaviour of ions (Chapter 2).

ION UPTAKE

The concentration of potassium ions in plant cell sap is about 10^{-3} M whereas in soil water it is about 10^{-5} M. This simple observation, that ions are present in plant tissues at higher concentration than in the liquid phase of the soil, raises three types of physiological question. These are concerned with:

(a) why, in metabolic terms, the ions are so concentrated;

Figure 1.5: A form of the soil–plant–air continuum concept which illustrates movements of ions and water

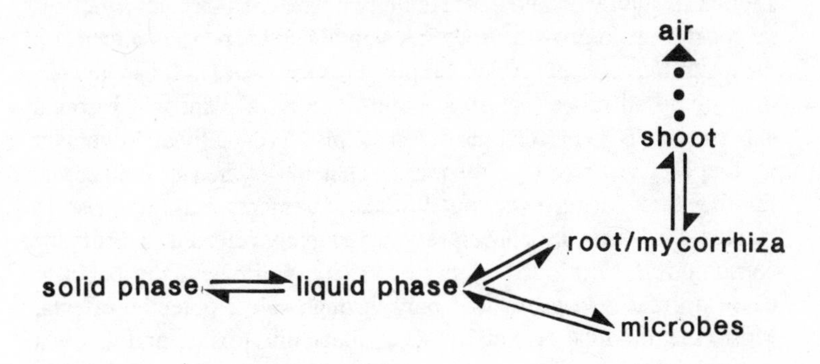

(b) how the penetration of the membrane system of the plant is achieved; and
(c) the significance of the tissue concentration observed.

The first two problem areas, the essentiality of elements and ion absorption, are still very active in research terms despite long histories of research and a voluminous literature. The interpretation of elemental concentrations in tissues is a subject in which enlightenment is badly needed despite wide use of tissue analysis in agriculture to diagnose mineral deficiencies. Data on this topic are also accumulating through application of the ecosystem approach to vegetation. Some trends can be discerned which serve as preliminary guidelines or warning signals for would-be interpreters.

For the ecologist it is worth outlining the salient features of each field and then reviewing briefly the metabolic roles of each element of interest. It is thought best to separate these reviews from the chapter on the 'availability' of elements (Chapter 11). Readers can then synthesise and integrate the two parts themselves.

ASSIMILATION AND UTILISATION OF IONS

Basic ground rules include the principle of limiting factors, the law of the minimum, and criteria of essentiality. Thus: 'The level of plant production can be no greater than that permitted by the most limiting of essential growth factors.' This expresses an idea, applied to plant nutrient limitations to crop growth at the close of the nineteenth century, that formed a basis for the rational use of fertilisers. It is applicable in many ecological situations and may be applied to supply of nutrients, toxicity of ions or other environmental constraints to growth. In the present context it permits a focus on the quantitative aspect of ion supply. From Figure 1.6 it can be seen that, given all other factors are non-limiting, a plant will respond experimentally to increments of ion supply. Eventually an increment no longer gives rise to a response. That ion is then no longer the limiting factor to growth, but further increments may give rise to increases in tissue ion concentration frequently referred to as luxury consumption. Continued supply of the ion will eventually begin to cause decreases in growth, if only through solute-potential effects. Most essential ions eventually act as metabolic toxins, and this is a phenomenon of ecological importance in relation, for example, to saline soils, mine wastes or soils derived from serpentine rock (see Chapters 18, 14 and 19 respectively).

The 'criteria of essentiality' concern the question 'How do we know that a particular ion is essential for growth of plants or for that matter any type of organism?' The problem arises because plant tissue contains a very wide range of elements. How do we test whether the presence of an element in tissue means that it is essential? The criteria for essentiality were set out in the course of an investigation into the role of copper as a trace element. In their short paper, Arnon and Stout (1939) proposed that an element cannot be considered essential unless:

(a) a deficiency makes it impossible for the plant to complete the vegetative or reproductive stage of its life cycle;
(b) such deficiency is specific to the element in question, and can be prevented or corrected only by supplying this element;
(c) the element is directly involved in the nutrition of the plant, quite apart from its possible effects in correcting some unfavourable microbiological or chemical condition of the soil or other culture medium.

Although formulated for application to physiological and agricultural research, these ideas should still have relevance for ecologists concerned with soil–plant or ecosystem effects. We can no longer concentrate only on the plant itself, as it is becoming clear that strict elemental requirements of plants are not the same as those of animals or microorganisms. Three examples will illustrate this view.

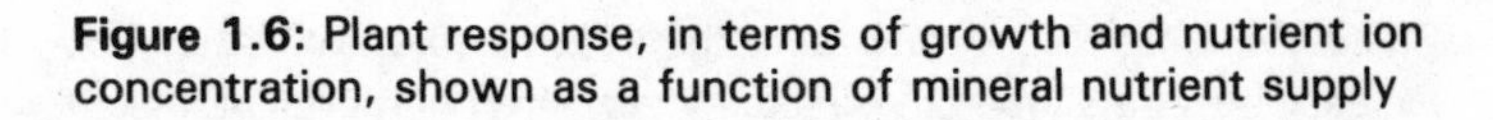
Figure 1.6: Plant response, in terms of growth and nutrient ion concentration, shown as a function of mineral nutrient supply

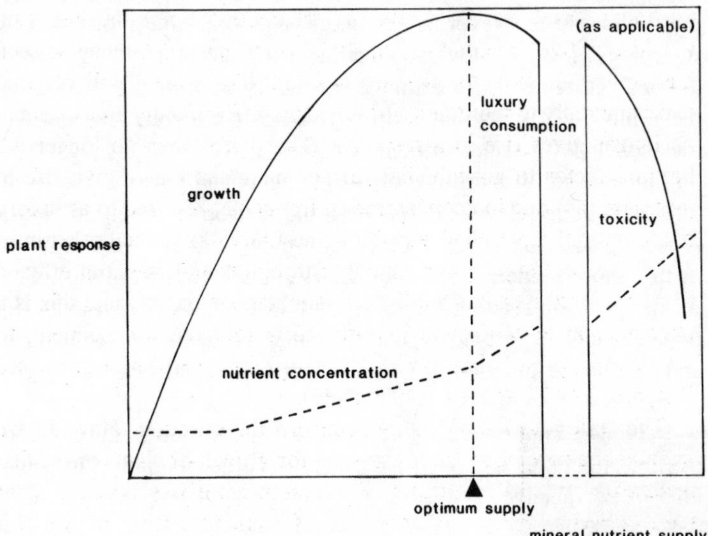

It has not yet been shown conclusively that cobalt is essential for plants in pure culture, but the nodulated legume system requires the element and so do ruminant animals (Young 1979). Plants do not absorb cobalt, and an interesting case of cattle becoming cobalt deficient occurred in Ireland when high soil manganese prevented cobalt uptake by vegetation (Poole *et al.* 1974). Selenium is also non-essential by the above criteria, but it is well known that the genus *Astragalus* (Leguminosae) can accumulate the element as seleno-cysteine and seleno-methionine. Selenium substitutes for sulphur in these compounds (Peterson 1971). The tissue concentrations available are sufficiently high to be toxic to stock, even though selenium is essential for the health of stock animals. It is suggested by Pate (1983) that selenium accumulation is one form of chemical protection against predation by invertebrate herbivores (see Chapter 5). A rather similar case can be made for silicon. Some groups, notably the grasses (Gramineae) and the horsetails (*Equisetum* spp.), accumulate silicon as silica. The plants can be grown in culture without silicate. Under conditions of tropical agriculture on highly leached soils, silicon deficiency leads to poor growth in sugar cane and reduced pest and disease resistance in rice (Sanchez 1976).

These examples indicate that there are ecological roles for elements beyond those envisaged by the strict 'criteria of essentiality' and testable by nutrient deficiency culture or purely metabolic studies. Nevertheless there is a place for metabolic studies in ecology. More ecological phenomena will eventually be resolved at a molecular level, for example the induction of nitrate reductase; the synthesis of metabolites protective against herbivores; and the metabolism of nutrient storage (Chapter 2).

The account that follows of the elements and ions of ecological interest does not dwell too much on the metabolic or physiological aspects, which are covered classically by Hewitt and Smith (1975) or in a more modern fashion by Clarkson and Hanson (1980). The former is worth reading for its account of methods of research into trace element requirements, and the latter for its attempts to explain and classify the functions of ions by their fundamental chemical properties.

It is also important to comment on assimilation pathways and possible toxicities as well as placing tissue concentration in some kind of perspective.

Absorption of ions

What does the physiological picture of ion absorption convey to the student of ecological systems? The following ideas seem to be useful, bearing in mind that experimental tissue-level physiology is practised on very simple systems over short time periods, and most whole-plant physiology is of greatest relevance to agriculture rather than to ecology. Recent reviews, especially those of Clarkson (1984) and Barber (1984), give balanced interpretations of these two areas of study.

(1) The boundary between the ionic environment and the living cell is the outer cell membrane system rather than the cell wall. Although ion exchange properties are assigned to cell walls, the walls of cells, except where suberised, are a pathway for water and ions (Figure 1.7).
(2) Ions accumulate across the cell membrane boundary into the cytoplasm–vacuole system. Because vacuoles occupy up to 90% of cell volume, vacuolar contents determine the ionic composition of the tissue.
(3) Ion absorption, and subsequent compartmentalisation, is an

Figure 1.7: Membrane systems in plant cells and the pathway for water and ion transfers

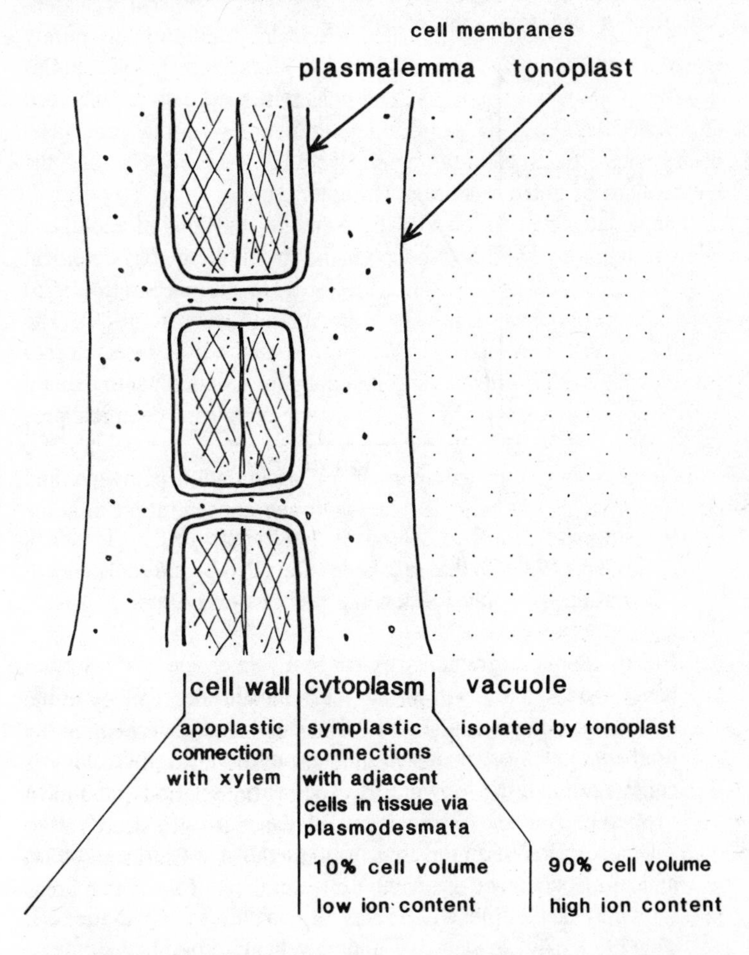

essential part of cellular metabolism. It has a number of characteristics:

(a) It is dependent on cellular energy.
(b) It entails enzyme-like carrier systems for traversing membranes. Such carriers are still largely hypothetical but explain why the relationship between rate of ion absorption by a tissue and the external ionic concentration is similar to the Michaelis relationship between enzyme and substrate (Figure 1.8).

Figure 1.8: The relationships between ion uptake rate and ion concentration to be expected if an enzyme-like carrier is responsible for penetration of cell membranes by ions

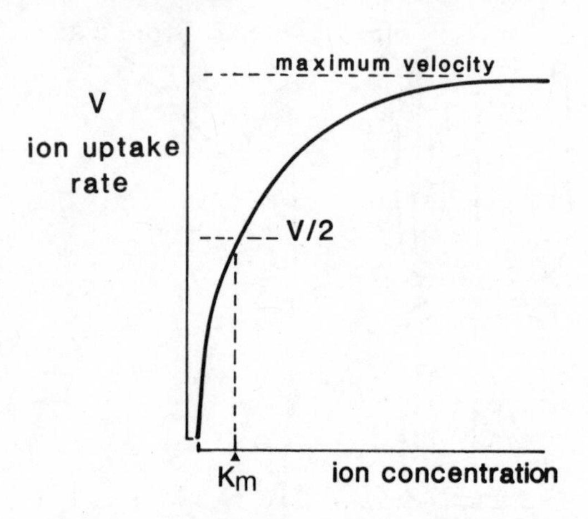

(c) Ions are absorbed selectively and controlled by feedback (Glass 1983) so that uptake is controlled by internal concentration, possibly influencing the carrier system.

(4) Electrochemical gradients exist. Many microelectrode studies have shown that, within the cell membrane, the vacuole–cytoplasm complex is slightly negatively charged. The charge gradient, in the order of 100 mV, means that electrochemically cations are moving 'down hill'. The charge separation is linked to metabolism and associated with the extrusion of H^+ ions. The absorption of anions thus occurs against both concentration and electrochemical gradients.
(5) Ions may be complexed on entry or soon after entry to the cell. This obviously steepens the ionic gradient across the cell membrane. In the case of phosphate ions, phosphate esters are formed very rapidly after assimilation and this is an entry point to phosphate metabolism. Complexing by specific short proteins, such as calmodulin for calcium (Dieter 1984) or the metallothioneins as complexing agents for copper or zinc, may either be a transport tactic or represent synthesis of a metabolically active entity. The surfaces of cells, and in turn the root surface, thus serve as sinks for ions with low or zero concentration in the solution close to the surface.
(6) Diffusion is one process supplying ions to the root surface. The

other process is the mass flow of water driven by evapotranspiration. Which is dominant or limiting will vary with each particular ion, its source and the environmental conditions.

2

Mineral composition of plant tissues and the function of ions

Chemical analysis of plant tissue will continue to be a prime means of interpreting soil–plant relationships. Tissue analysis has been used as a diagnostic tool in agriculture, and many of the drawbacks of the approach are well understood. Ecologists should be aware of the factors leading to variation in the elemental composition of tissues, which can be dealt with under three headings: technical, biological and environmental. Some topics will be expanded on in other sections, and this section should be regarded as little more than an annotated checklist.

TECHNICAL FACTORS

(1) Is the sample uncontaminated by soil or dust? Simple, but not infallible, checks are (a) to determine the ash content, and (b) to examine plant material directly or after wet ashing for silica particles of soil origin. Samples with ash contents greater than 5% should be treated with suspicion. Remember that soil contamination will both affect mineral composition and reduce the quantity of tissue weighed for analysis.

(2) Is the analytical procedure capable of determining the element in question? Modern analytical procedures for trace elements do need careful application in spite of their convenience to the non-chemist.

(3) Has the tissue been adequately sampled in terms of determining field heterogeneity?

(4) Is the sorting or homogenisation of plant material adequate? (See Chapter 12.)

BIOLOGICAL FACTORS

1. Seasonal factors

Seasonal changes are to be expected. These will not be predicted properly or explained until we have good mass balance studies for more plants, and more elements, that take into account all parts of the plant and that measure gains and losses of minerals.

Seasonal changes are illustrated for phosphate, nitrogen and magnesium for temperate zone perennial grasses, a subtropical perennial grass and an annual chenopod halophyte (Figure 2.1). In the case of the perennial grasses, above-ground tissues only are included. In our analysis of *Salicornia*, root tissue is included.

The pattern for nitrogen and phosphorus seems to have three possible biological components:

(a) a period in which high levels of supply from underground storage organs or seeds are supplying a low biomass;
(b) a period of sustained growth and biomass increase when concomitant nutrient uptake is not sufficient to sustain highest concentrations;
(c) a period of retranslocation to storage organs or senescence and loss of elements.

The example of magnesium shows that the pattern for two grasses can be different. *Agrostis stolonifera* tissue has a virtually constant magnesium concentration, whereas in cultivated *Lolium perenne* the concentration increases to the end of the growing season before declining. Two other patterns were described by Tyler (1971) for plants of a Baltic seashore meadow. Silicon accumulated steadily through the season in above-ground tissues. This was especially marked in the grass *Agrostis stolonifera* and in *Eleocharis* (Cyperaceae). A group of elements including iron, aluminium, nickel and lead exhibited high concentrations in spring and autumn but low in midsummer. These patterns are summarised diagrammatically in Figure 2.2. The problem with these data is that underground parts are not being considered, and the fluxes of elements are not clear.

Figure 2.1: Seasonal patterns of concentration of ions in plant tissue. (a) Leaf tissue phosphate in three grasses and a halophyte: *Lolium perenne* (fertilised pasture); *Agrostis stolonifera* (Baltic seashore meadow); *Hyparrhenia rufa* (subtropical savannah); *Salicornia dolichostachya* (Chenopodiaceae, intertidal mudflat). (b) Leaf tissue nitrogen in the above species. (c) Leaf tissue magnesium for two species. (Data from Tyler 1971; Daubenmire 1972; Flemming 1973; B. Madden and D.W. Jeffrey, unpublished.)

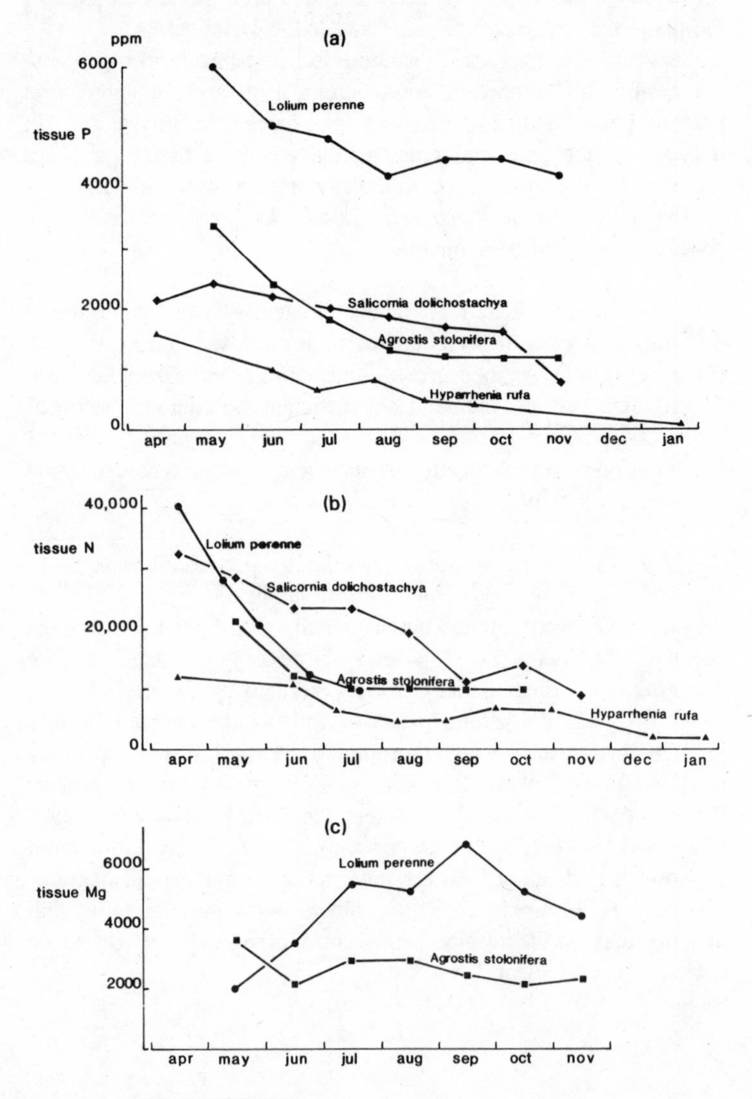

Figure 2.2: Seasonal changes in concentrations of minor elements in above-ground tissues of *Agrostis stolonifera* (parts per million) from Baltic seashore meadow. Note differences in concentration scale. (Data of Tyler 1971.)

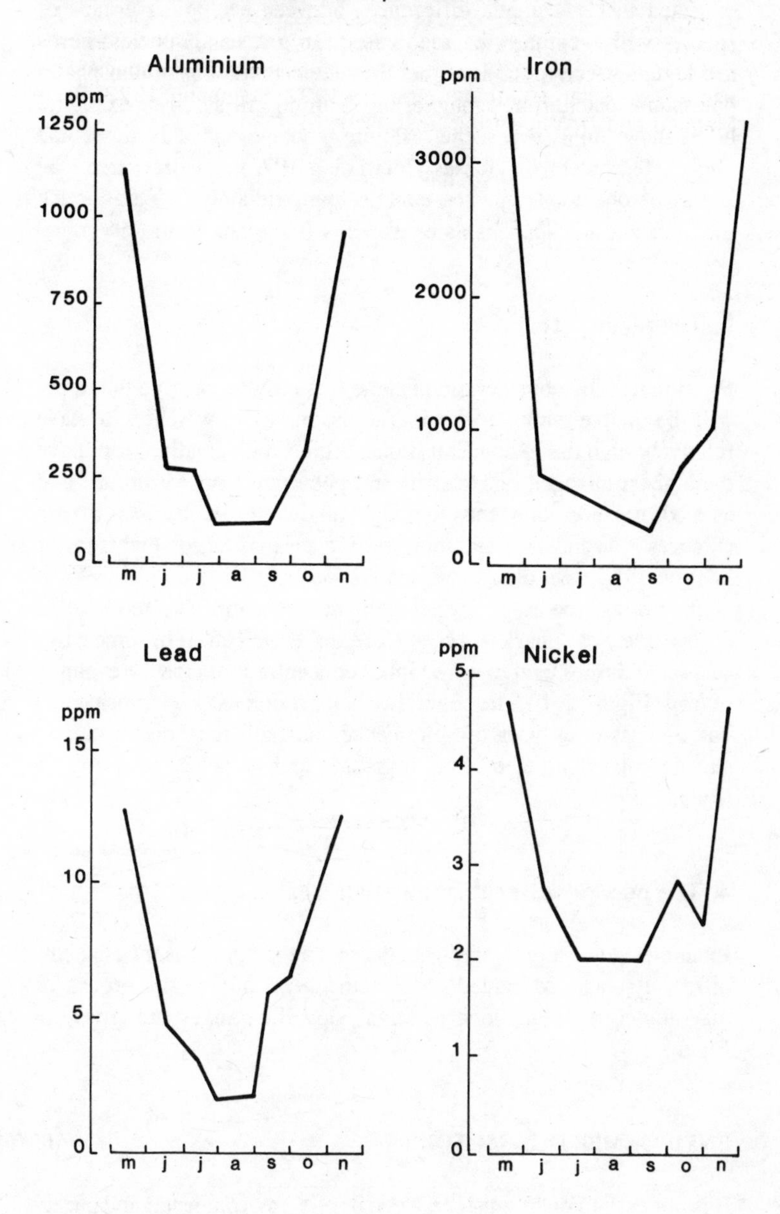

2. The plant species

Differences between species have been referred to already, and it is expected that systematic differences between ecological groups of species will eventually be recognised. In grassland species, herbs and legumes tend towards higher tissue concentrations of phosphate, potassium, calcium and magnesium than do grasses. For example, herbs have four times the calcium content and 1.5 times the phosphate content of grasses (Flemming 1973). In forest trees the leaves of deciduous species contain approximately twice as much nitrogen and phosphorus as evergreens (Cole and Rapp 1981).

3. The plant part

From metabolic considerations alone it is easy to imagine that there will be active sinks for mineral assimilation which will have relatively high tissue concentrations. Almost universally, seeds have the highest nitrogen phosphorus and potassium concentrations, and as seeds develop, this tends to weight analysis of the flowering parts (Figures 2.3a and b). Calcium or iron levels may be somewhat lower in flowering parts than in vegetative issues.

In woody species mineral element concentrations tend to be ordered: leaves > bark > wood (Cole and Rapp 1981). In some cases senescent tissues tend to have higher concentrations (see the example of lead, Figure 2.4). The generality is for minerals to be translocated out of tissues as part of senescence metabolism. This is clearly shown, for example, by tundra grasses and sedges as discussed in Chapter 17.

4. The mineral element in question

Examples given already indicate that metabolic patterns for elements differ. It must be added that elements also vary in orders of magnitude of tissue concentration. Specific ranges are given in Figure 2.5.

ENVIRONMENTAL FACTORS

It is not easy to distinguish between purely environmental and purely

Figure 2.3: (a) Nitrogen and (b) phosphorus deployment in oat plants (*Avena sativa*) throughout the growing season. Data replotted as a percentage from Williams (1955)

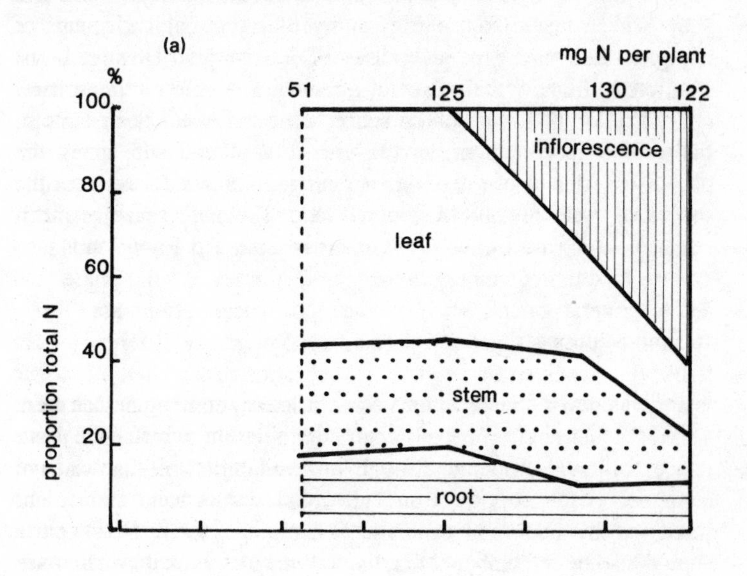

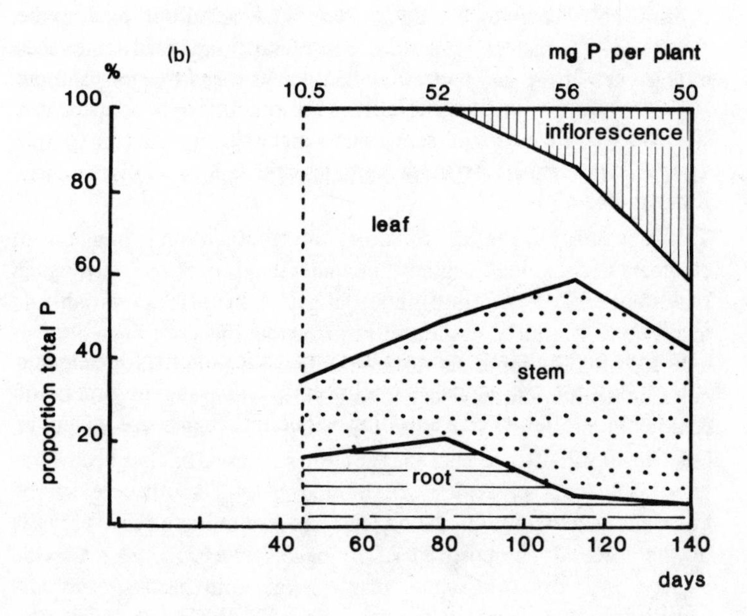

biological causes for particular tissue concentrations. The balance between ion accumulation and the 'dilution' of absorbed ions by tissue growth illustrates this. As the section on the availability of ions will show, mechanisms controlling supply of a particular element are complex and may generate pulses of ion supply. However, one example will show the powerful effect of a simple factor, namely concentration of a solid-phase source of a non-essential and potentially toxic trace element, lead (Figure 2.4). The results stem from one of a series of pot experiments carried out with a system fully described in Jeffrey and Maybury (1981). The salient feature of this experiment is that a grass, *Cynosurus cristatus*, is grown in a sand culture medium containing increasing quantities of solid-phase lead as the mineral galena, PbS. Tissue lead concentrations are clearly related to substrate, green tissues responding rather differently from non-living senesced tissue. The concentration of lead in dying tissue is a well known phenomenon which lacks a metabolic explanation.

Most other environmental factors that are said to influence tissue concentration also operate through an 'availability' mechanism. For example, tissue copper, iron and cobalt in herbage tend to be increased by poor soil drainage (Flemming 1973). However, it should also be remembered that the volume of soil explored by roots may decrease as soils become anaerobic from below. This may place a limit on absorption of these ions. The relationships between environmental factors, soil solid phase and liquid phase and plant biology are at the heart of many ecological research programmes.

METABOLIC FUNCTION OF ELEMENTS

This account of the assimilation, metabolism and function of elements is compiled to support an understanding of their ecological importance. The literature underlying this material is variable in quality and quantity, and a uniform treatment has not been attempted. It is more important to summarise the main ideas where a voluminous literature exists and to indicate signposts or gaps where ecological needs are not met. As a guide to the proportions of elements normally present in plant tissues, Figure 2.5 has been compiled. The data presented are limited to two sources reviewing grassland vegetation. On the one hand is Flemming's (1973) review of the mineral composition of herbage, which includes a wide variety of information on elements, grass and herb species and agronomic situations. On the other are the data on the European

Figure 2.4: Tissue lead concentrations of the grass *Cynosurus cristatus* grown experimentally in a sand culture containing finely ground galena (PbS). (D.W. Jeffrey and M. Maybury, unpublished.)

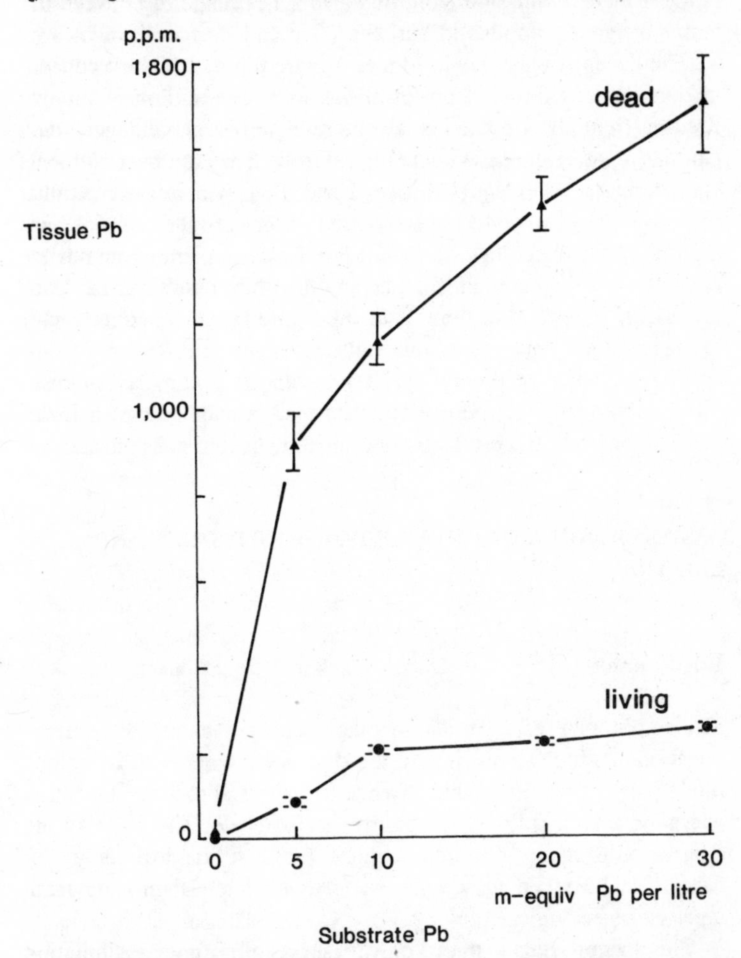

meadow grassland sites studied in the International Biological Programme (Titlyanova and Bazilevitch 1979). Here data from the above-ground green tissue were utilised. These two data sets, when combined to illustrate ranges, do so more usefully in this context than the encyclopaedic compilations of other authors, e.g. Bowen (1979). In particular, the latter does not give a balanced perspective including data from extreme situations, for example the sodium content of halophytes or the metal content of accumulator species. Two comments may be made. First, silicon values may be higher than the norm because of the inclusion of many grasses. Secondly, selenium is the principal omission from this data set; a range of between 0.1 and 1.0 ppm Se applies to herbage (Ure and Barrow 1982).

The group A elements in Figure 2.5 are those of highest concentration range, expressed uniformly as parts per million of the dry tissue weight. It is not especially helpful to use percentages when comparing one element with another. Group B are the trace elements considered to be essential to plants and their symbionts. Note that the two ranges extend over seven orders of magnitude. Group C are a group of elements that are commonly found in plant tissue but are not known to be essential for plants. Most other possible additions to this list would be in the 0.1 to 0.01 ppm range. A commentary is made on the contents of this figure in Table 2.1.

The elements will now be dealt with in groups as follows: nitrogen and sulphur; phosphorus; potassium and sodium; calcium and magnesium; trace metals; and silicon, boron and chlorine.

ASSIMILATION AND METABOLISM OF NITROGEN AND SULPHUR

Introduction

The importance of nitrogen in plant ecology cannot be overemphasised. Its importance lies in the discrepancy between its unique biochemical roles, its virtual absence from crustal rocks and the high energetic cost of splitting the dinitrogen molecule. There is a strong temptation to make 'nitrogen ecology' a rich subfield of study, but care must be taken to weave the nitrogen strands into the main tapestry of ecological biology.

This section deals with ecological aspects of nitrogen assimilation and metabolism. Nitrogen fixation ecology will be treated separately as background and in case-history form (see Chapter 4). Sulphur is included because of similarities in assimilation and the close metabolic relationship of the two elements.

Assimilation of nitrogen

Nitrogen may be assimilated by plants from solution in two forms,

Figure 2.5: Mineral composition of plants from grassland sites. Compiled from data of Flemming (1973) and Titlyanova and Bazilevitch (1979). See explanation and commentary in text

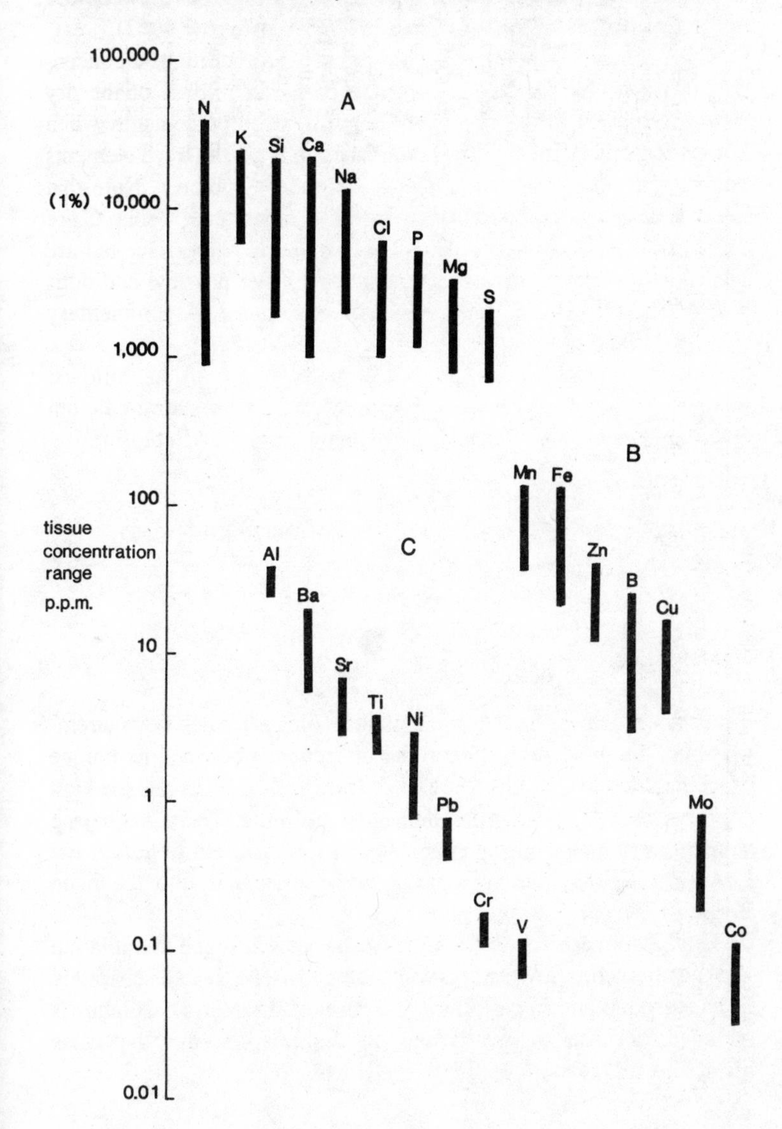

Table 2.1: A commentary on elements occurring in plant tissues (see Figure 2.5)

Element	Ecology			Biochemistry/physiology		Agriculture		Notes on ecology
	Major	Restricted	Accumulated[a]	Basal metabolism	Secondary metabolism	Major	Local	
N	+	+		+	+	+		Of very great and all-pervasive importance. Ecology is many faceted, with interest in patterns of assimilation, rate of turnover, secondary metabolites and nitrogen fixation.
K				+		+		Basal metabolic needs are met by environment except in cropped systems.
Si		+	+		?		+	Significance in terrestrial ecosystems probably not sufficiently understood
Ca	+		+	+		+		Ca interacts with many other elements metabolically, facilitating uptake or protecting against toxicity. Differentiation of calcareous and non-calcareous habitats is a major ecological phenomenon.

Na	+			+		+		Saline soils are main focus of interest
Cl	+			+	+	+		Saline soils are main focus of interest
P	+	+		+		+		Second only to N as a major limiting factor to growth. Assimilation of environmental P through mycorrhizal systems is a major evolutionary response
Mg		+		+				Serpentine and coastal soils
S		+		+	+	+		In conjunction with N occurs in many secondary metabolites of ecological interest
Mn		+		+			+	Toxic in poorly aerated acid soils
Fe		+		+			+	Deficiency to some species at high pH, may be toxic in conditions of poor aeration
Zn		+	+	+			+	Toxicity known on geological anomalies and industrial wastes
B		+		+			+	Toxicity on reclaimed fuel ash and in salt deserts
Cu		+	+	+			+	Toxicity known on geological anomalies and industrial sites

Table 2.1: *Continued*

Element	Ecology			Biochemistry/physiology		Agriculture		Notes on ecology
	Major	Restricted	Accumulated[a]	Basal metabolism	Secondary metabolism	Major	Local	
Mo				+			+	Metabolic needs appear to be met by environment
Co		+	+	+			+	Importance is in context of N fixation and animal nutrition. Toxicity possible on serpentine
Al		+	+				+	Mobilisation in acid soils leads to community differentiation through toxicity/tolerance
Ba			+					Low solubility of sulphate removes potential toxicity
Sr		+	+					^{90}Sr accumulation of radioecological interest
Ti								Apparently inert and often used as an internal reference element in plant analysis
Ni		+	+					Toxicity documented on anomalies, e.g. serpentine
Pb		+					+	Industrial pollution and metal anomalies lead to toxicity

Table 2.1: *continued*

Cr	+	+	Toxicity encountered in industrial wastes and on serpentine
V		+	

[a] Also F, I, Se, Sn, W, Rh, Hg, Ni (Peterson 1971).

Explanation

Ecology

Major: element influences community distribution and differentiation on a large scale. Ecological effects may be of the same order as those resulting from sunlight, temperature and rainfall
Restricted: influences either on a small scale in landscape terms or related to the ecology of particular taxa
Accumulation: the element may be accumulated by particular taxa, producing tissue concentrations at least ten times the concentration indicated. Data from Peterson (1971).

Biochemistry/physiology

Basal metabolism: element is shown to be involved in a key biochemical process and by inference is assumed to be of universal importance.
Secondary metabolism: a universal role is not inferred, but the role of the element may be of importance to particular species.

Agriculture

Major: the principal elements routinely used in crop fertilisers.
Local: elements of interest either as fertiliser additives or in view of toxicity.

nitrate (NO_3^-) or ammonium (NH_4^+) (Figure 2.6). In most soils nitrate is the predominant ion, and the current view of its assimilation involves a series of enzyme-mediated chemical reductions. These reduction steps require trace-element cofactors.

An alternative pathway to the GS-GOGAT system (see Figure 2.6) which was once thought to be of prime importance, is through the amination of α-oxoglutarate by the enzyme glutamate dehyrogenase (GDH). This system appears to have a lower affinity for ammonium than GS-GOGAT, and Pate (1983) suggests that it functions as a back-up system.

There are many possible features of ecological importance in the assimilation of nitrogen, quite apart from those associated with nitrogen fixation, which are dealt with in Chapter 4. Eventually it may be anticipated that a clear pattern will emerge, but at present we are dealing with a set of fragmentary reports. For example, there are three kinds of variation possible with respect to the nitrate reductase system: (a) its behaviour with respect to soil nitrate; (b) its location in the plant; and (c) if nitrate is universally acceptable as a nitrogen source.

Estimation of nitrate reductase activities has been used by G.R. Stewart in the interpretation of the ecology of temperate saltmarshes (see Chapter 18), and of several West African plant communities (Stewart and Orebamjo 1983). Figure 2.7 is compiled from data presented in this work and illustrates the frequency of nitrate reductase activity levels estimated in leaves of grasses, forbs, shrubs and trees from rainforest and savannah communities. The original data of activity levels in species have been simply classified and the frequency of each class is displayed as a bar chart.

The forb-rich successional stage to rainforest indicates a high frequency of high enzyme levels, a situation not observed elsewhere. Although a preponderance of low enzyme activity is common, the woody communities have a mixture of both high and low activities (Figure 2.7a). All the savannah communities had characteristically low nitrate reductase activities, with only damp savannah species demonstrating any elevated activity. The communities were further characterised by an experiment in which nitrogen fertiliser was applied to the vegetation (Figure 2.6b). Re-estimation of nitrate reductase activity showed that the nitrate-assimilating potential of communities was only close to fulfilment in the forb-rich fallow community. The lowest level of potential appears to apply to savannah woodland and mangrove-swamp communities. This work provides a strong strand of evidence towards showing that supply of

Figure 2.6: A scheme for the assimilation of ammonium and nitrate by plants

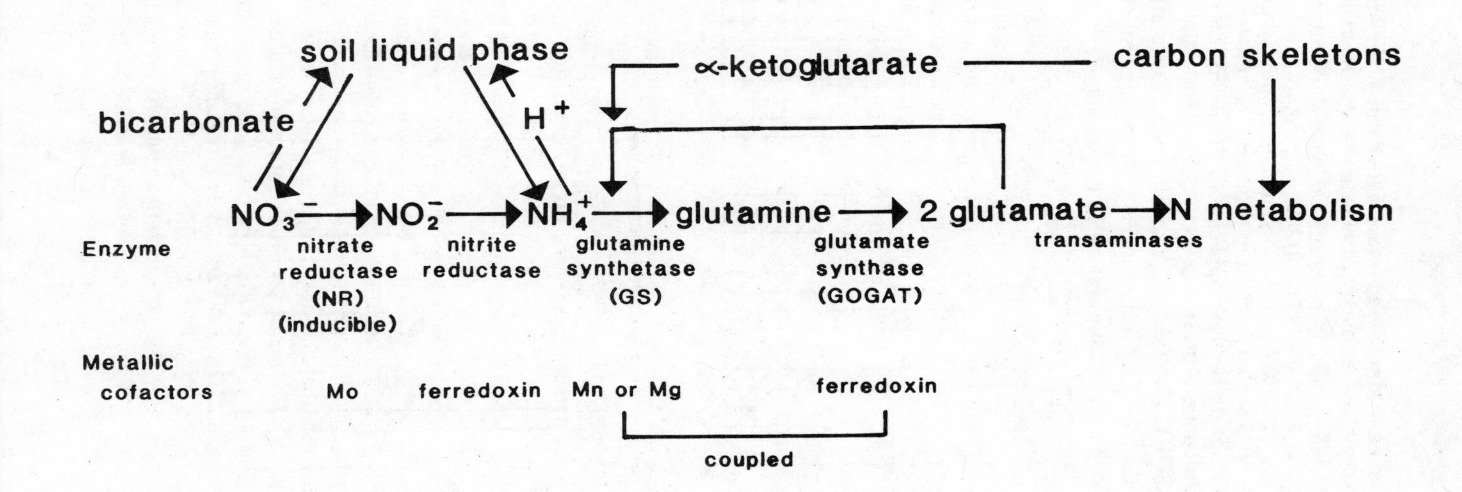

Figure 2.7: Studies of nitrate reductase activity in West African plant communities. (a) Frequency of nitrate reductase activity levels in leaf tissue taken from species in the field. Numbers in parentheses are species examined in each community. (b) Nitrate reductase activity potential in various communities. Activity before fertiliser application is expressed as a percentage of activity after nitrate fertiliser application. Compiled from Stewart and Orebamjo (1983)

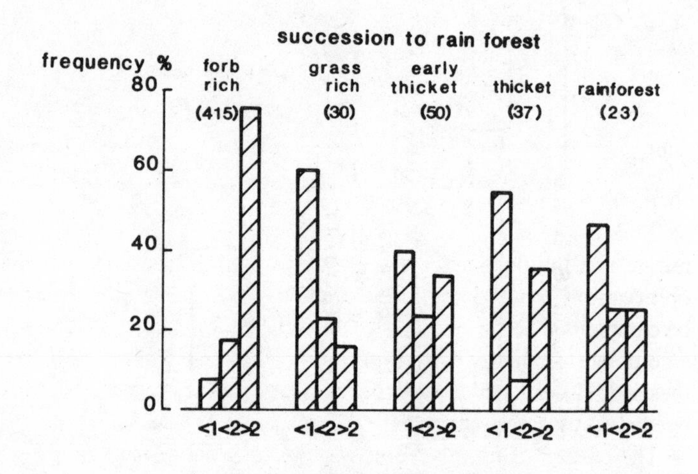

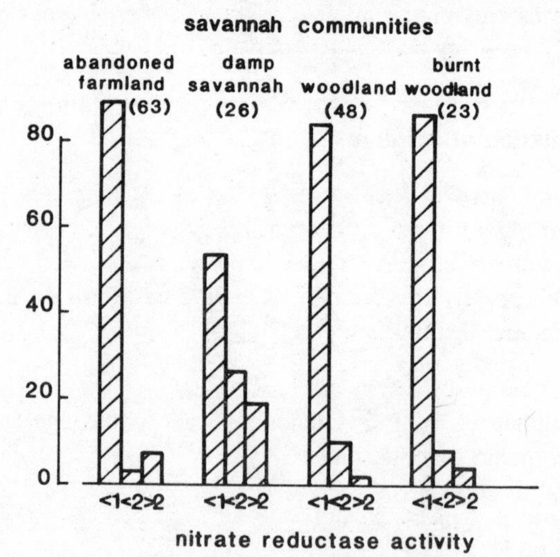

(a)

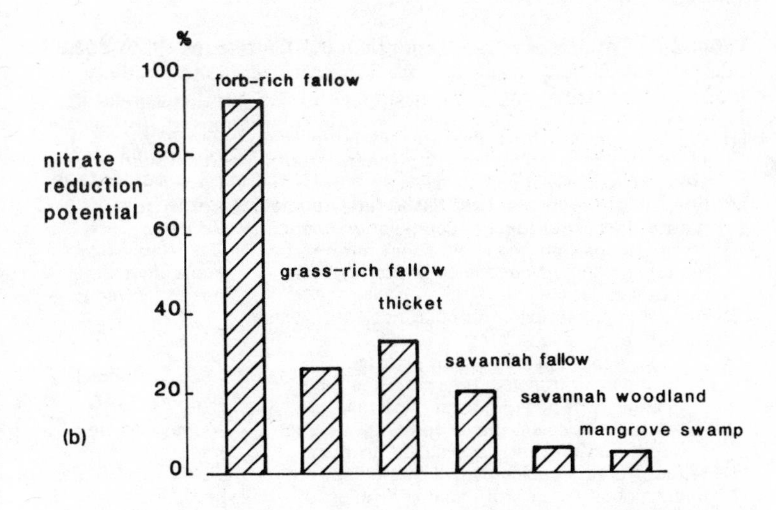

nitrate is a limiting factor to vegetation growth in the field. These observations, on 103 rainforest and 110 savannah species, need to be complemented by systematic soil-nitrification studies, and ideally a complete quantitative view of the nitrogen cycle needs to be synthesised. However, the example indicates the sheer quantity of work needed to assemble this preliminary view.

The pattern of nitrate reduction in ecological groupings of plant species as part of the broader picture of nitrogen assimilation has recently been reviewed by Pate (1983) (Table 2.2).

Assimilation of sulphur

Sulphur is absorbed and translocated as sulphate ion, but must be reduced before it can play a metabolic role as a constituent of proteins or other active molecules. The reduction steps will not be as comprehensively covered here as in the case of nitrate reduction, but the principal features are:

(1) Activation of sulphate:
Sulphate + ATP → adenosine 5-phosphosulphate (APS) + inorganic pyrophosphate
(2) Stabilisation of phosphate–sulphate linkage:
APS + ATP → adenosine-3-phosphate-5-phosphosulphate (PAPS) + ADP
(3) Transfer to a carrier (Car) with at least one thiol group:
Car–SH + PAPS → Car–S–S–OH + PAP

Table 2.2: Pattern of nitrate reductase (NR) in tissues (Pate 1983)

		Examples
(1)	Annual species with a shoot-dominated pattern of nitrate assimilation. Nitrate present in root xylem, especially if soil nitrate high	*Xanthium*, *Gossypium* *Cucumis*
(2)	Root nitrate reducers. Leaf tissue rarely contains nitrate, but under luxury accumulation shoot nitrate is possible. Inducible shoot nitrate reductase, limited directly to photosynthesis, can be activated	*Hordeum*, *Triticum*, *Zea*, *Raphanus*, apple (*Pyrus*), peach (*Prunus*)
(3)	Annual legumes fixing nitrogen but with access to nitrate:	
	(a) temperate species which produce amides as translocatory products have active root-based nitrate reductase	*Pisum*, *Vicia*, *Lupinus*
	(b) tropical species, ureide forming, are shoot reducers if nitrate is available to them	*Vigna*, *Glycine max*
(4)	Woody forest species. Not well studied, but forest communities, for reasons discussed elsewhere, may need to assimilate ammonium rather than nitrate. Stewart and Orebamjo's (1983) data already discussed are relevant in showing that woody species from rainforest and savannah woodland have inducible NR in leaf tissues and relatively high field levels of activity	*Celtis* spp., *Ficus* spp., *Trema guineensis*, *Musanga cecropioides*
(5)	Woody species in which nitrate reductase activity appears to be virtually absent and non-inducible	Ericaceae

(4) Reduction with ferredoxin:

$$\text{Car–S–S–OH} \xrightarrow{\text{Reduced Ferredoxin Oxidised}} \text{Car–S–SH}$$

Dithiol carrier

(5) Transfer of outer thiol group to acetyl serine to form cysteine:

Dithiol carrier + acetylserine → cysteine + carrier + acetate

(6) Utilisation of cysteine in metabolism

Utilisation of nitrogen and sulphur

Proteins contain about 16% nitrogen and up to 4% sulphur; hence crude protein is often estimated as total N × 6.25. Protein is the general currency of nitrogen and sulphur transfer in ecosystems, and hence has a central role in all aspects of biology that can be taken for granted here.

Within plants nitrogen is being continuously transferred between growing and senescing organs. The compounds involved in translocation vary with species (Pate 1983) but are generally nitrogen-rich amides or ureides. It can be anticipated that species may have characteristic nitrogen flow patterns on a diurnal and seasonal basis. It is clear that senescing leaves lose nitrogen which is either reutilised immediately if vegetative growth is active, or stored if fruiting or dormancy have been initiated. This is also reflected in Figure 2.1 which shows a systematic decline in nitrogen concentrations of vegetative tissues throughout a growing season. The enrichment of seeds with stored protein and other essential elements, in addition to carbohydrate, is a universal phenomenon of great ecological significance. Young seedlings are independent of external ion sources for periods normally extending from a few days to perhaps some weeks. The seed is also an obvious target for herbivores, leading, in evolutionary terms, to protective tactics adopted by species.

Nitrogen- and sulphur-containing secondary metabolites of ecological importance

It is probable that phenolic compounds are the largest group of substances concerned in either anti-predator tactics or in allelopathy. (Allelopathy means biochemical interference between one organism and another: see Rice 1984.) However, nitrogen- and sulphur-containing compounds are also conspicuous and common in these roles (Figure 2.8). Nevertheless, it is not always possible to interpret clearly the function of a specific substance. For example, many alkaloids or cyanohydrins with toxic or other physiological effects are known to be seed-germination inhibitors (Rice 1984), but Bernays (1983) discusses their effectiveness in reducing insect attack. Commonly known alkaloids in this group include cocaine, caffeine, quinine, strychnine, codeine, ephedrine and atrophine. Sorghum seedlings contain a cyanogenic glucoside, dhurrin, with enzymes which hydrolyse it to glucose, hydrogen cyanide and parahydroxybenzaldehyde. The potential of this type of reaction in discouraging predators has been demonstrated in *Sorghum, Lotus, Acacia, Prunus* and *Manihot esculenta* (cassava). Most reactions of this kind which liberate hydrogen cyanide do so when tissues are damaged. Tissue damage also triggers the formation of volatile and offensive di- or polysulphides from *Allium* species, or mustard oils from members of the Cruciferae (Robinson 1983). The role of selenium-containing amino acids as protective toxins has already been referred to, but there are other non-protein amino acids with the same role. The protein-

Figure 2.8: Metabolites with roles in protecting seeds and other plant tissues against herbivores. (a) Nitrogen-containing antagonistic metabolites and the generation of free cyanide. (b) Polysulphides, seleno amino acids and toxic amino acids

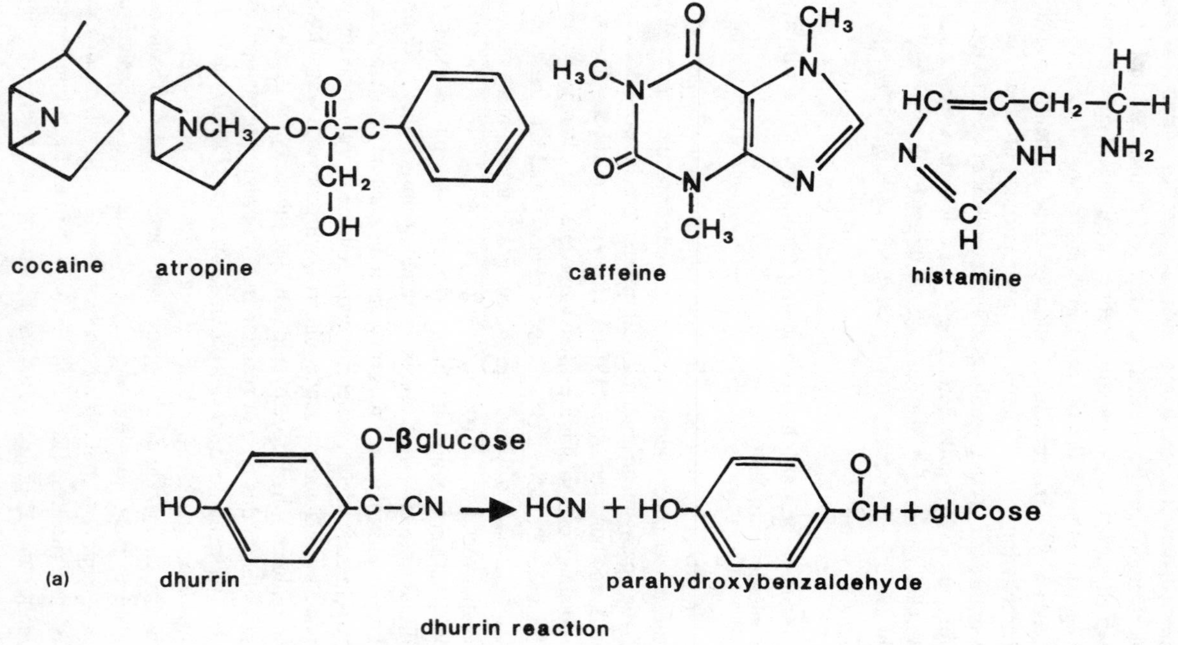

cocaine

atropine

caffeine

histamine

(a)

dhurrin

$HO-C_6H_4-C(O\text{-}\beta glucose)-CN \rightarrow HCN + HO-C_6H_4-CHO + glucose$

parahydroxybenzaldehyde

dhurrin reaction

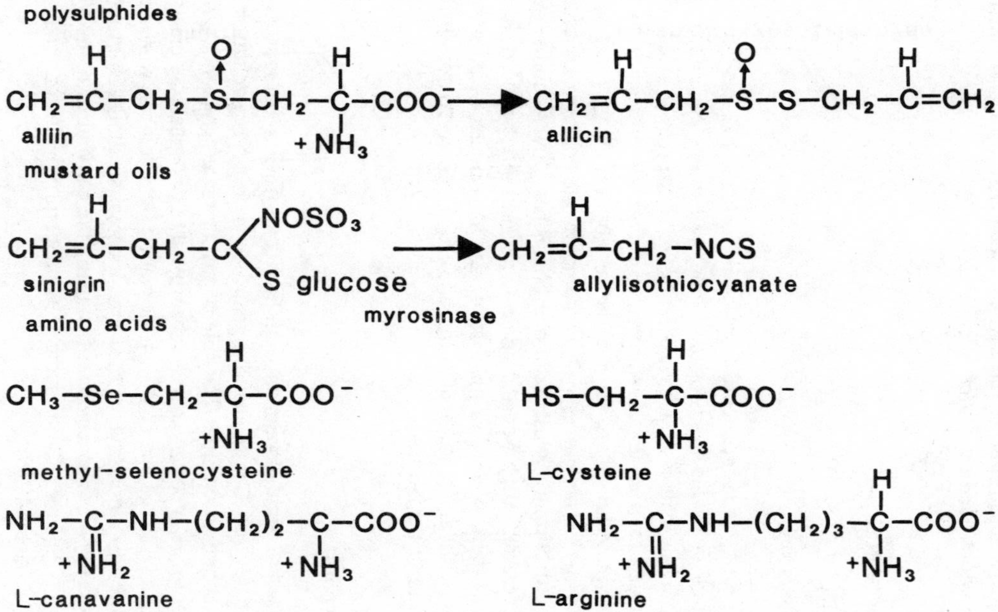

polysulphides

$$CH_2{=}CH{-}CH_2{-}S(\rightarrow O){-}CH_2{-}CH(NH_3^+){-}COO^- \longrightarrow CH_2{=}CH{-}CH_2{-}S(\rightarrow O){-}S{-}CH_2{-}CH{=}CH_2$$

alliin — allicin

mustard oils

$$CH_2{=}CH{-}CH_2{-}C(NOSO_3)(S\ \text{glucose}) \xrightarrow{\text{myrosinase}} CH_2{=}CH{-}CH_2{-}NCS$$

sinigrin — allylisothiocyanate

amino acids

$$CH_3{-}Se{-}CH_2{-}CH(NH_3^+){-}COO^-$$

methyl-selenocysteine

$$HS{-}CH_2{-}CH(NH_3^+){-}COO^-$$

L-cysteine

$$NH_2{-}C({=}NH_2^+){-}NH{-}(CH_2)_2{-}C(NH_3^+){-}COO^-$$

L-canavanine

$$NH_2{-}C({=}NH_2^+){-}NH{-}(CH_2)_3{-}CH(NH_3^+){-}COO^-$$

L-arginine

(b)

rich seeds of the Leguminosae are commonly protected by direct neurotoxins or amino acid analogues. The best known is L-canavanine, which is an analogue of arginine. Seeds are also protected by a number of glycoproteins (lectins) which function as haemagglutinins. They form between 2 and 10% of the protein of seeds. The red strains of the kidney bean *Phaseolus vulgaris* may contain enough of this toxin to cause serious human health problems. This lectin is water soluble and may be removed by soaking for 24 hours followed by thorough cooking in fresh water. The stinging hairs of *Urtica dioica* contain the amine histamine. This is obviously a defence against some predators, but not against a range of butterfly larvae which feed on *Urtica* specifically.

PHOSPHORUS

Introduction

Phosphorus, as an element of prime importance to plant ecology, runs a close second to nitrogen. Again we have an element, in the form of phosphate ions, which is central to all aspects of metabolism and structure from the genetic structure of all organisms to the massive bones, teeth and tusks of the vertebrates. On the other hand, phosphorus is not an especially common element in the crustal rocks; furthermore, most metal phosphates have low solubilities and high densities. High demand and low accessibility have led, in my view, to the evolution of plant strategies as complex as those concerned with the capture of nitrogen. See Chapter 16 for further discussion of this idea.

Absorption

Because of high dispersion and limited diffusion of sources of phosphate in soil, high absorbing surface area is a key requirement for effective absorption. The role of mycorrhizal symbionts is thus of very great importance in the absorption of phosphate by plants in the field (see Chapter 4). Most plant physiological work has, however, used non-mycorrhizal young cereal seedlings as material for short uptake experiments, often using the convenient radioactive isotope ^{32}P. From these studies two general conclusions may be drawn (Pitman 1976):

(1) The absorption sites operating at the low phosphate concentrations obviously involved in the field depend on calcium ions and NADPH for effective absorption.
(2) Of the phosphate entering root tissue, most of it is rapidly incorporated into organic compounds, especially sugar phosphates.

In mycorrhizal systems and the few Australian heath species investigated with ^{32}P, long-chain polyphosphates may also be rapidly synthesised (Jeffrey 1968; Cox *et al.* 1975). Whatever the precise molecular mechanism of uptake, the phosphate ion is quickly removed, maintaining the 'zero sink' at the cell wall concerned. Hence the maximum concentration gradient is maintained into the soil system.

Phosphatase activity is also reported from root surfaces collected from grasses (Woolhouse 1969) and *Pinus radiata* (Theodorou 1971). The significance of these observations is not clear but it opens the possibility of the otherwise inert inositol hexaphosphates of soil being utilised as a phosphate source by mycorrhizal plants (see Chapter 4).

Assimilation

Labelled phosphate is incorporated into metabolic systems as rapidly or slowly as the state of growth of the plant demands. Sinks for phosphate incorporation in all plants include meristems, expanding leaves and developing fruits. These appear to be 'spared' from deficiency under low-supply conditions. For example, a cereal under extreme phosphate deficiency will continue to develop a new leaf as one senesces prematurely. Similarly, a few seeds will be produced in each inflorescence, rather than none. The view given by Williams (1955) of phosphate deployment in oat (Figure 2.3b) would seem to be applicable to many annual plants. In the seed phosphate is stored as the calcium or magnesium salts of inositol phosphates (Figure 2.9). These compounds have great chemical stability but may be hydrolysed by phytases which become active on seed germination. What is not clear is the role of other storage organs as sinks for phosphate during the growing season. Underground plant parts probably store phosphate remobilised on senescence during periods of dormancy in perennials. Prime subjects for study include species with a high phosphate demand, such as *Urtica dioica*, and species surviving on phosphate-limiting substrates — tundra and bog graminoides, for example.

A new tool for the investigation of this problem is beginning to bear published results, namely ^{31}P nuclear magnetic resonance

Figure 2.9: Polyphosphates and inositol hexaphosphate, which serve as phosphate storage compounds

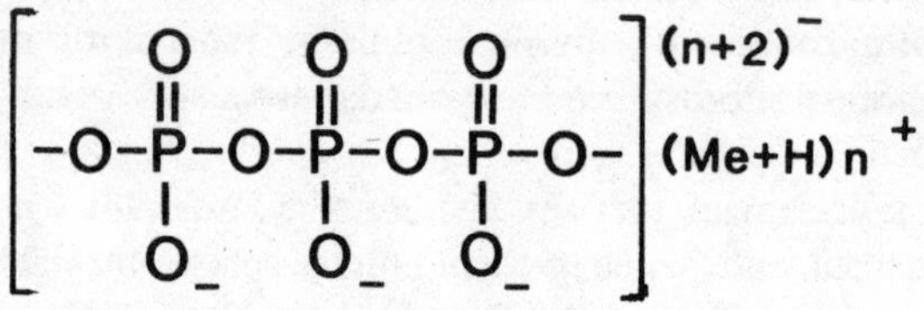

polyphosphate

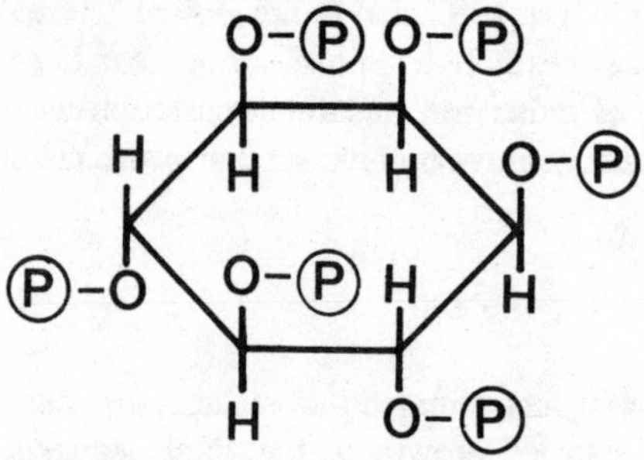

inositol hexaphosphate

(NMR). This technique permits the identification of phosphate-containing molecules either in extracts or in living tissue. It is essentially a non-destructive method and may be combined with other types of analysis. Its potential value in ecological studies is the rapid determination of a spectrum of compounds, permitting intercomparisons.

POTASSIUM AND SODIUM

Absorption

In most roots there is a preferential absorption of potassium compared with sodium. Both these common unilavent ions will move along the electrochemical gradients of tissues, but because of discrimination at cell membranes, or sodium extrusion, the ultimate concentration ratio may be 20 K^+ to 1 Na^+.

Assimilation and utilisation

Both ions participate in determining the solute potential of cell vacuoles, potassium normally being of prime importance. However, one appears to substitute for the other in this role. Sodium influx to the vacuole may play a major role in permitting turgor maintenance under saline environmental conditions. Sodium efflux is also an important process, eliminating sodium from the cytoplasm where it is clearly toxic, even in halophytes. This efflux occurs across the plasmalemma (Pitman 1976). Potassium is the predominant cation in the ion flux responsible for effecting stomatal opening.

The activation of a few enzymes is known to depend on a univalent cation, and in the plant enzymes studied *in vitro* potassium or sodium are interchangeable (Hewitt and Smith 1975). Because of both technical difficulties in totally excluding sodium and the lack of co-ordinate bonding of either ion, it is not possible to state with confidence that sodium is universally essential in the classic sense.

CALCIUM AND MAGNESIUM

Introduction

These two common divalent cations behave very differently within plant tissues. Generally, calcium is combined in complexes with low turnover, whereas magnesium is more mobile. In neither case should we imagine that their metabolic or ecological importance has been fully explored.

Absorption

Absorption of calcium by roots seems to be confined to the youngest parts of the roots and it moves to the xylem via the apoplast. There is strong compartmentalisation within cells, with low cytoplasmic and relatively high vacuolar concentrations. The controlling mechanism here is in dispute (Kirkby and Pilbeam 1984).

Magnesium accumulation by whole plants is strongly dependent on linkages with energy metabolism, and there is some evidence that the 'younger root only' uptake pattern is also followed. Within the plant, distribution is strongly towards the symplast according to demand. Here it is distributed between cytoplasm and vacuole.

Magnesium can be the second most important vacuolar cation after potassium. Should potassium supply be limited, magnesium concentrations may rise.

Assimilation of calcium

Most calcium in a plant is held exchangeably into the cell wall structure as cation bridges between galacturonic and glucuronic acid molecules. Removal of this calcium with a complexing agent, e.g. ethylene diamine tetraaectic acid (EDTA), leads to softening of the cell wall. More calcium is associated with fixed sites on cell membranes, especially the exterior of the plasmalemma, stabilising them and facilitating their function. Again, the ion forms bridges between phosphate and carboxylate groups of phospholipids and protein molecules at the membrane surface. Under conditions of calcium deficiency and senescence, membrane disorder is evident. The presence of calcium seems to offer a degree of protection against toxic metals which may be related to membrane integrity.

Calcium is involved in the activation of some membrane-bound enzymes but its most important activating function may lie in the 'calmodulin system'. Calmodulin is a specific short protein which reversibly binds Ca^{2+} with great specificity and affinity. The calcium–calmodulin complex has been shown to be the activating agent in a number of biochemically important systems.

The literature on calmodulin, as reviewed by Dieter (1984), is less than five years old and it is not yet possible to evaluate its full significance. The potential significance is that this system may offer a solution to some of the many problems of cell responses to external stimuli. These include the phenomena of cytoplasmic streaming, pollen-tube growth and *Mimosa pudica* leaf movement. These are all phenomena requiring calcium ions. A set of important processes in animal metabolism are also calmodulin-mediated, and the plant molecular biology seems similar where investigated.

Assimilation of magnesium

The most obvious role for magnesium is as a part of the chlorophyll molecule, where it is complexed as a tetrapyrrole ring. About 10% of magnesium in a leaf may be accounted for as chlorophyll. The characteristic intervenal chlorosis of crop plants with magnesium deficiency seems to be a direct consequence. Other chloroses, caused by iron deficiency or heavy metal toxicity, are more related to interference with iron-dependent tetrapyrrole formation.

Magnesium has four other major roles: as a binding agent in cell ultrastructural integrity, as an activator of enzymes, as a counterion for H^+ exchanges and as an osmotically important vacuolar ion. This is a large area of basal metabolic and physiological importance, creating a substantial demand in tissues. It is generally met by environmental supply in all but the most intensive cropping systems. Within the plant, magnesium is moderately mobile in contrast to calcium.

High tissue concentrations of calcium and magnesium

Solid-phase inclusions of calcium oxalate are occasionally noted, for example in the Caryophyllaceae. Alternatively, a soluble form of sequestered calcium, calcium malate, may be found in vacuoles. These are interpreted as a means of removing excess calcium (Rorison and Robinson 1984). Similarly, calcium carbonate may be precipitated on leaf epidermal surfaces via glands in the Saxifragaceae or in the cell walls of the Cruciferae.

High tissue concentrations of magnesium are known to be toxic to many species. This is regarded as one component of the phenomenon of the 'serpentine' floras of the world (Proctor and Woodell 1975) (Chapter 19). The precise mechanism of magnesium toxicity is not yet elucidated. Explanations may eventually include effects related to high Mg/Ca ionic ratios and toxic effects of the magnesium ion alone. Species that are tolerant to high environmental magensium may control uptake, tolerate high tissue concentrations or even have a higher than normal requirement for the element.

TRACE METALS: Fe, Mn, Zn and Cu, Mo, Co

Iron

Plant iron metabolism is concerned with the properties of a large variety of iron-containing molecules with reducing functions. The two main groups are the compounds in which Fe^{3+} is held in a porphyrin ring, and the iron proteins.

The haem compounds include all the cytochromes, peroxidase, catalase and the haemoglobin of legume nodules. The main group of non-haem iron metabolites are the ferredoxins, but the nitrogenase complex of bacteria contains iron as well as molybdenum and magnesium.

The metabolic importance of iron, and its tendency to remain in the cell once elaborated into a molecule, mean that a growing plant has a continuing demand for the element. The low solubility of iron in aerated soil of near neutral pH is a problem overcome by two devices. Reducing activity of plant roots means that iron is absorbed as Fe^{2+}, ferrous iron, and transported as a chelate with citrate. Roots may also lower the pH of soil solution adjacent to them, which would have the effect of increasing the solubility of many ions.

In soils with low redox conditions, massive concentrations of soluble ferrous iron may be present in soil leading to toxicity to some species and tolerance in others (see Chapter 9).

Manganese

The behaviour of manganese in soil resembles that of iron, but in the plant manganese resembles calcium and magnesium. It is involved in the activation of enzymes in a manner to some extent replacing magnesium. Toxicity of reduced manganese is possible in plants growing in soils with low redox potential. The intensity of this effect is difficult to distinguish from that of ferrous iron.

Zinc and copper

Unlike iron and manganese, zinc and copper are not normally abundant in soil. It is assumed that these ions are actively accumulated, and there is some evidence that transport is in chelated form. In mammals specific cysteine-rich metal binding proteins, metallothioneins, are responsible for controlling the ionic concentration of these metals in tissues (Vallee and Ulmer 1972). The protein concentration is related in an inducible manner to metal levels. This is one of the mechanisms that may be invoked to explain the tolerance of species and ecotypes to high environmental zinc and copper (see Chapter 14). Evidence is starting to accumulate that illustrates the presence of plant metallothioneins and their role in binding copper, zinc and cadmium. However, other copper-binding proteins, with less cysteine, have been isolated from plant tissue (Tukendorf and Baszynski 1985).

Enzymic zinc proteins exhibit a mixture of strong covalent bonding of the metal and a situation permitting removal with a chelating agent. In the latter, another metal can be substituted. In contrast, all copper enzymes are strongly covalently bonded.

Molybdenum

Molybdenum as molybdate is the heaviest essential element and the one required at lowest tissue concentration. It is essential for the fixation of dinitrogen by nitrogenase systems and the reduction of nitrate. Nitrogen-fixing systems generally have the higher Mo requirement, and that might conceivably be a limiting factor in the field.

Cobalt

As explained earlier, cobalt is included because of its essentiality to the symbiotic bacteria and actinomycetes concerned with nitrogen fixation. The cobalt atom is chelated in the nucleus of coenzyme B_{12}. This factor is required for the production of haem compounds essential in the biochemistry of nitrogenase.

CHLORINE AS CHLORIDE, SILICON AS SILICATE, BORON AS BORATE

Chloride

There is an interesting ambivalence of view between the biochemists and physiologists about the importance of the chloride ion. On the one hand biochemists point out that very small concentrations of chloride ion will suffice for water-culture-grown plants to prevent metabolic disorders. Its function in basal metabolism is associated with the evolution of gaseous oxygen in photosynthesis. A few organochlorine compounds are also noted in the biochemical literature (Hewitt and Smith 1975).

On the other hand, physiological experimentation and analyses for ecological purposes reveal that not only is chloride ion actively absorbed (Pitman 1976), but that it may also constitute up to 6000 p.p.m. of tissue dry weight in grassland plants (Figure 2.3). Chloride, together with sulphate and hydroxyl ions, is commonly a major vacuolar anion. It tends to be excluded from cytoplasm, where organic anions permit electrochemical balance for cations.

Silicate

Silicate, along with borate, appears to be absorbed passively in quantities related to transpiration. However, it must be admitted that technical difficulties in analysis, combined with only a small degree of economic interest, have left these elements less than well studied.

Opal phytoliths, which are small (<5 μm in diameter) grains of pure amorphous silica, SiO_2, are found in the tissues of many plants. Monocotyledons contain 10–20 times as much Si as dicotyledons. These may be sufficiently characteristic to be of value in palaeoecology or forensic investigations, as their structure is preserved in ash. The manner in which an opalith forms in a cell or in which cell walls are silicified is not understood. Because of the high silica content of cereals used in brewing, beer is virtually a saturated solution of silicate!

Borate

The requirements of plants for boron are very low. In water culture they may be satisfied by ions diffusing from a borosilicate glass container. It is not surprising that we still know few details of its metabolic role. It is undoubtedly involved with carbohydrate transport and the biosynthesis of lignin. Boron deficiency leads to a variety of tissue malformations and may be of local agricultural importance. Toxic concentrations of borate have been encountered in substrates such as pulverised fuel ash.

SUMMARY AND CONCLUSION

Mineral nutrient elements play particular roles in metabolism. Their ecological importance is dependent on these roles, evironmental supply and ultimately relationships in ecosystems.

In summary their metabolic roles are:

- in the coordinate bonding of nitrogen, phosphate and sulphur in the molecular architecture of all organisms
- in the ionic contributions to cell turgor
- in ionic binding of cellular structures

— as ions in the maintenance of enzymic configuration
— as functional groups of enzymes
— in the relationship between plants and predators
— as ions that exert a toxic effect

It can be seen that the perspective of the ecologist differs from those of the biochemist/physiologist and agronomist. It is emphasised that much of our knowledge of mineral nutrition is gained from agriculturally motivated studies. Future ecological studies will continue to utilise the tools of agricultural research, including use of stable and radioactive isotopes, modern forms of electrochemical and spectrographic analysis, ^{31}P nuclear magnetic resonance and a range of plant-culture and microscopic techniques. By extending the range of species studied for ecological purposes, the generalisations we make about 'basal mineral metabolism' will be strengthened. Undoubtedly, new discoveries will be made about deployment of mineral nutrients in taxa with particular ecological characteristics.

3

Plants and water

THE LINK WITH PHYSIOLOGY AND BIOCHEMISTRY

The major thrusts of water-related plant physiology are concerned with the strong interactions with plant production. These emphasise the reciprocal exchange of water and carbon dioxide at the wetted mesophyll cell wall and the exchange control exerted by stomatal behaviour. Metabolic economy of water use is also well studied, with the C−4 metabolic pathway for photosynthesis now being interpreted as a familar feature of tropical and arid-zone species. The 'crassulacean acid metabolism' (CAM) pathway is an even more extreme metabolic device for conservation of water. It permits photosynthetic energy capture during the day to be separated from gas exchange at night. Obligate water loss from open stomata is thus minimised, and the stored water of succulents conserved. This system is known to operate not only in the Crassulaceae, but also in succulent members of the Cactaceae, Aizoaceae, Asclepiadaceae, Bromeliaceae, Orchidaceae and Liliaceae.

THE SOIL–PLANT–AIR CONTINUUM (SPAC)

The idea of water potential is one key to the understanding of the soil–plant continuum, to which the atmosphere must be added. If you are not familiar with this system, choose a text such as Kramer (1983), Salisbury and Ross (1985) or Meidner and Sheriff (1976) to learn the fundamentals. These tell us that, by observing the free energy of water in different parts of a system, it is possible to predict its behaviour quantitatively. By resolving the ways in which free energy increases or decreases and unifying the unit of measurement (a pressure unit 0.987 atmospheres = 1 bar = 10 MPa), it is possible to directly compare the binding (reduction in free energy) of water in various systems.

Vascular plants may be described as a set of cellular osmotic systems, living cytoplasm and vacuole, in contact with a series of porous wetted spaces, cell walls and lumens of dead cells (Figure 3.1, and see Figure 1.5). Movement of water within the plant, and from soil to air via the plant, is driven by gradients in water potential. The value of water potential, ψ, and its algebraic sign (+ or −), represents the extent to which the free energy of water is altered. Pure water at standard temperature and pressure is conventionally given the value of zero MPa. Compression or heating add energy and give a positive value. Interactions of water with solutes, with a matrix of surfaces, or by exposure to less than standard pressure lead to reduced free energy and negative values. In plants, water potential can be resolved into two predominant components. ψ_s, solute or osmotic potential, is always negative and ordinarily covers the range −1.0 to −2.5 MPa. It is generated by organic and inorganic solutes in cell cytoplasm and vacuoles. Photosynthesis and ion absorption are the basis for metabolic control of solute potential and absorption of water. Pressure potential, ψ_p, may be generated either in turgid cells (positive) or in water columns under tension in xylem (negative). The phenomenon of root pressure may also result in a transient positive ψ_p in xylem.

Probably in all living systems there is a ψ_m, matric potential, component. This would arise from the microfibrillar cellulose matrix of cell walls and from interaction with large colloidal molecules. In practice this is not readily separable from ψ_s, and is seldom referred to in the literature. Its magnitude is probably quite low, i.e. < 0.1 MPa. In some non-vascular plant systems — algae and lichens especially — large extracellular pools of polysaccharide may represent a matric potential component of some size. Imbibition of seeds may also entail the saturation of a matric reservoir prior to metabolically linked water absorption.

If metabolic processes give rise to water absorption, the sustained flow of water through plants and vegetation, evapotranspiration, is caused by the water potential difference between soil and air. This is commonly a differential in the order of 100 MPa. The differential results from the input of long-wave solar energy heating soil and plant surfaces and surrounding air (see Chapter 9).

Figure 3.1: Diagram to illustrate the pathway of water and ion movement through the soil–plant–air system

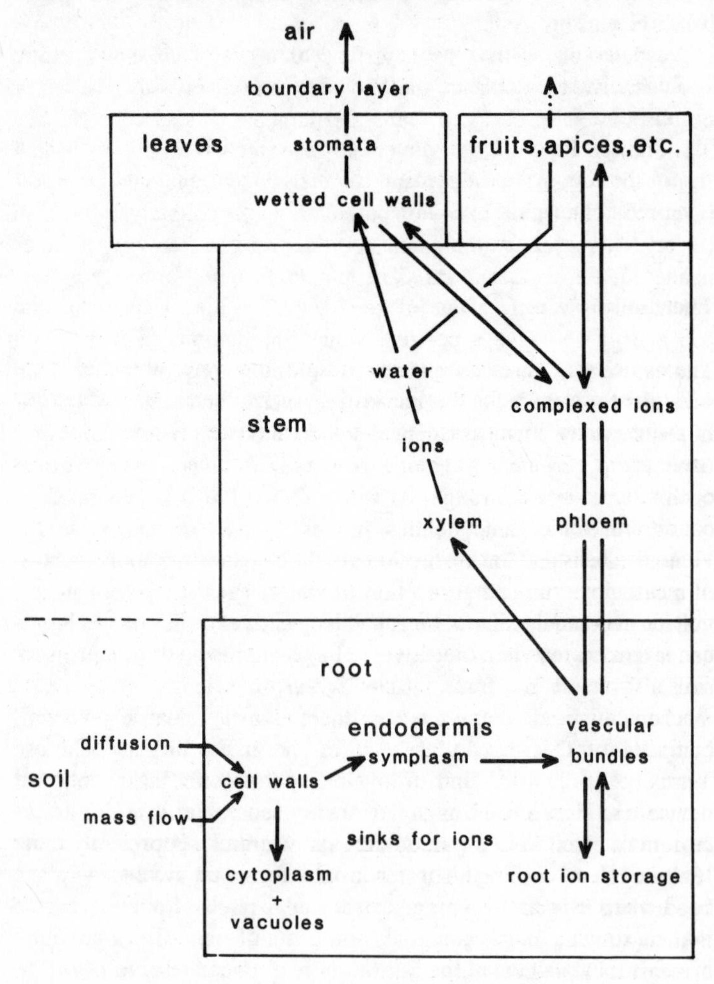

WILTING

Wilting of vegetation means that the pressure potential of tissues is relatively low, usually because water loss is greater than uptake. This may be due to a number of circumstances:

(a) Rapid transpirational loss by leaves is not matched because of internal resistance to flow. This is probably a common diurnal phenomenon in bright sunlight.

(b) Soil moisture is low and ψ_m soil is more negative than (ψ_s + ψ_p) tissues. This may lead to permanent wilting and is the usual situation defined on the soil-moisture characteristic curve (Figure 8.8).

(c) Transpirational loss cannot be replaced because of deficiencies in the root system including freezing, anaerobic conditions and predator or pathogen damage.

Thermocouple psychrometer

The device that gives the greatest insight into water potential in the soil–plant system is the thermocouple psychrometer. Its main virtue is the capacity to measure total water potential in small samples (milligrams of water) quite rapidly (several minutes). Although the equipment is portable, it is not easily used in the open under field conditions, for reasons which will become obvious.

At the heart of the device is a miniature thermocouple, capable of measuring small temperature differences and generating a microvoltage when cooled. This is mounted in a sample chamber which can take many forms, including a ceramic chamber for insertion into soils. Typically a sample is placed inside, the chamber is sealed mechanically and the sample then equilibrates with the relative humidity of the very small chamber air volume (Figure 3.2a). Temperature equilibration must also be complete all through the device and during each phase of the measurement cycle. A small current is then passed briefly through the thermocouple to cause 'Peltier effect' cooling. A water droplet forms on the thermocouple bead when it is cooled below dewpoint (Figure 3.2b). This droplet is then allowed to re-evaporate, and evaporative cooling generates a measurable microvoltage in the thermocouple (Figure 3.2c). A standard curve may be constructed linking water potential of standard solutions with the microvoltage. Molal solutions of sodium chloride are usually employed as standards.

It is important to realise that the linkage between sample and microvoltage is as follows:

Sample ψ → equilibrium RH of air → evaporation rate → cooling effect → microvoltage output.

The device is very susceptible to two influences, temperature

Figure 3.2: The dewpoint psychrometer sample chamber illustrating stages in estimation of water potential of a sample

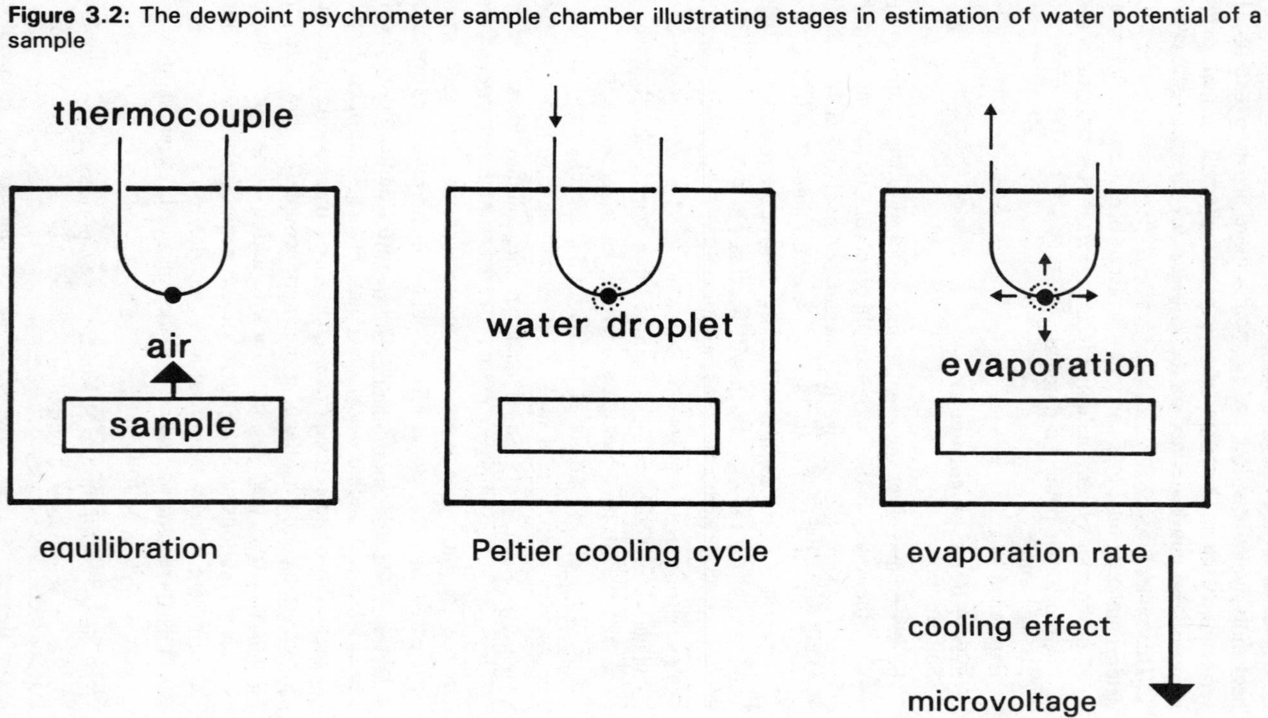

fluctuations and contamination by perspiration, dust, saline solutions, etc. A sample chamber needs protection from draughts, body heat and any other thermal influences. The system must also remain scrupulously clean, to make sure that the sample chamber, and above all the thermocouple bead, remain uncontaminated.

This system is suitable for the following types of water potential determination:

	Water potential determined
Tissue samples	$\psi_p + \psi_s$ cells
Expressed sap	ψ_s vacuole
Soil slices	$\psi_m + \psi_s$
Soil solution	ψ_s
Emplaced in soil	$\psi_m + \psi_s$

WATER ECONOMY OF PLANTS

Plant evolution has come to terms with an interesting design paradox. How does a water-filled device, designed to intercept solar energy and provide a sink for carbon dioxide diffusion, avoid desiccation and excess heat load? Analysis of this problem needs to consider a number of parts of the SPAC including:

— interception of water
— secure water storage
— control of water loss
— desiccation tolerance
— water stress tolerance

A series of adaptive modifications are usually employed by species capable of surviving environmental aridity. These are illustrated by the four types in Figure 3.3. Not depicted are the ultimate drought avoiders, annual plants. These require reliable physiological means for detecting the commencement of the wet season.

WATER STORAGE

Most fully turgid plants carry a small amount of water, stored in vacuoles, which may be regarded as a short-lived buffer against losses greater than inputs. However, for plants capable of surviving seasonal aridity, one measure is long-term tissue water storage.

Figure 3.3: Four perennial plant types characteristic of arid zones

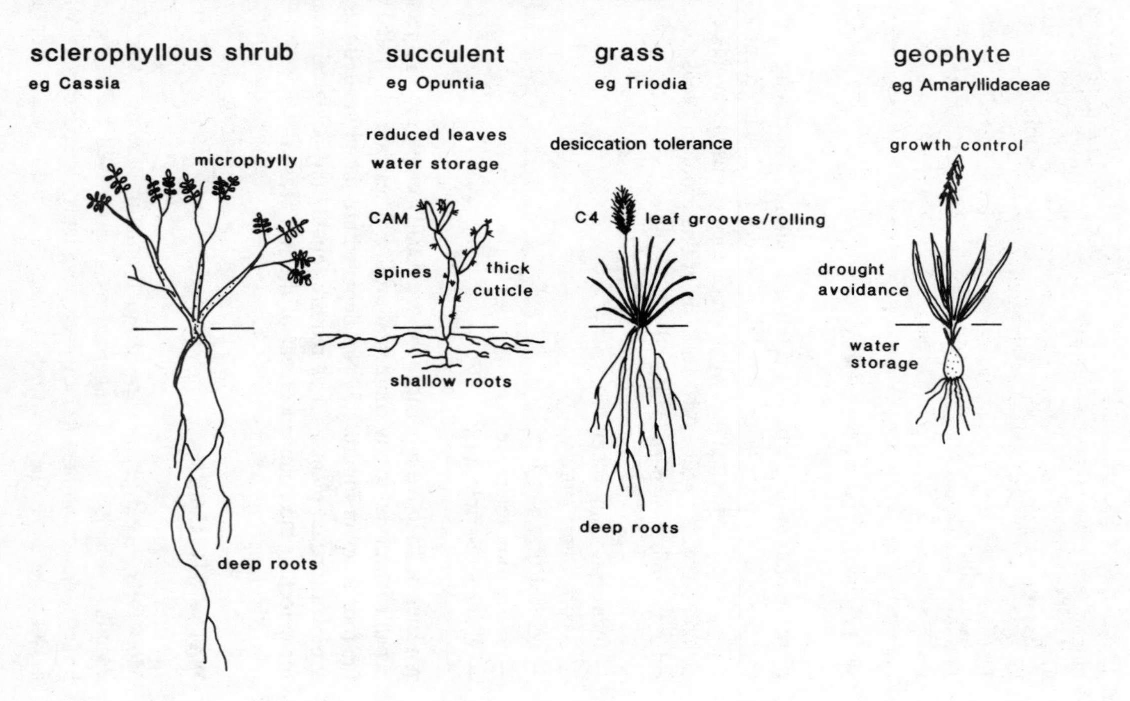

Arid-zone succulent plants, to be distinguished from succulent halophytes, have developed in a few major families, sometimes with prolific speciation and endemism. The many morphological variants make such genera attractive house plants.

Table 3.1 indicates the major families with succulent water storage tissues. Where investigated, the tissue has been found to be parenchymatous and has a solute potential in the order of 0.05 MPa. This demonstrates a clear difference from halophytes in which succulent cells have high ion content and low solute potential (Chapter 18).

It is inferred that the very elaborate spines of some types, e.g. Cactaceae and Euphorbiaceae, have the prime purpose of reducing predation of the water content. This simple view of the armour needs careful review, first because of the elaborate behavioural and anatomical specialisations in arid-zone faunas, and secondly because of other possible functions for the spines. These functions include the stabilising of boundary layers, reflection of long-wave radiation, and the trapping of occult precipitation. Another protective feature is the possession of allelochemic defences, which are widely present, for example, in the Euphorbiaceae and Asclepiadaceae. Many of the types named possess a range of other adaptations to the habitat, which complement storage.

WATER INTERCEPTION

For interception of soil water, two tactics are employed: (a) seeking groundwater; and (b) maximising interception of surface water. In both cases it should be remembered that soil water will only move under gravity but that rate of movement will depend on capillary size. In relatively dry soils only fine capillaries will contain water, and thus movement, even downwards, will be slow. Root growth rate will probably exceed rate of water movement. In some groups, deep rooting is very common, with various prairie and dune grasses, e.g. *Agropyron smithii* (wheatgrass) and *Ammophila arenaria* (marram grass), well known for fibrous root production at depths greater than 1 m. Persistent tap roots in some species may reach greater depths, but the literature is anecdotal rather than systematic. Deep rooting is an essential feature of sclerophyllous-leaved trees and shrubs characteristic of mediterranean-type climates.

In general, arid-zone succulents tend to have a framework of permanent suberised roots close to the surface which grow to

Table 3.1: Plant families with well developed water storage morphology

Family (No. of genera)	Representative genera	Notes on genera
Liliaceae (250)	*Aloe*, > 270 spp.	Small medium or tree-like with rosettes of thick leaves. Leaves are persistent in contrast to drought avoiders in family. Africa
Agavaceae (20)	*Agave*, > 300 spp.	Wide range of size, long-lived and with armoured leaf tips. Often monocarpic. Americas
Dioscoridaceae (5)	*Testudinaria*, 6 spp.	Slow-growing, long-lived yam with cork-covered tuberous root and vine-like shoots. South Africa
Cactaceae (50 conservatively)	*Carnegiea*, 1 sp.	The giant saguaro, the largest succulent, 15 m high, 200 y age, distributive limit Arizona–Mexico
	Opuntia, > 300 spp.	Prickly pear and cholla. Heavily armoured group with a complex ecology. Americas
	Mammillaria, > 250 spp.	Large group of small species. Mexico
Crassulaceae (35)	*Crassula*, *c.* 200 spp.	Range of morphology, unarmoured, mainly southern Africa
	Sedum, > 500 spp.	Small plants, wide range of morphology, ecological range tropics to subarctic in northern hemisphere
Aizoaceae (130)	*Greenovia*, 4 spp.	Rosette forms, endemic to Canary Islands
	Lithops, 75 spp. *Pleiospilos*, 33 spp.	'Living stones'. Plants almost buried in stony desert. South and south-west Africa
Euphorbiaceae (300)	*Euphorbia*, *c.* 350 succulent spp.	Wide range of form from arborescent to small barrel types. May be heavily armoured with thorns. All species yield latex which may be distasteful to predators. Africa
Portulacaceae (19)	*Anacampseros*, 55 spp.	Small plants with fleshy roots. Southern hemisphere

Table 3.1: Cont.

Asclepiadaceae (130)	*Stapelia,* > 100 spp.	Very reduced leaves and succulent stems. Africa
	Ceropegia, 44 spp.	Small shrubs and climbers. Central and southern Africa, Canary Islands, India
Compositae (900 but few genera succulent)	*Senecio,* 36 succulent spp.	All the succulent species of this large cosmopolitan genus are from South Africa
	Kleinia, 44 spp.	Close to *Senecio,* but distributed more widely. Jointed stems with deciduous leaves in some species
Geraniaceae (5)	*Pelargonium,* 24 succulent spp.	Shrubs with succulent stems. Deciduous. Southern Africa

produce absorbing roots during the season of rainfall. Even suberised roots may absorb water via discontinuities such as lenticels (Caldwell 1976). Spacing of arid zone species by various means is also a possible mechanism for optimising water interception and stored soil water. *Larrea divaricata* is shown to have a spacing related to such a simple moisture index as total rainfall (Figure 3.4).

Two properties of above-ground surfaces are possibly important in water interception. In rosette species, especially those with waxy surfaces, e.g. *Echeveria* (Crassulaceae) and *Agave* (Agavaceae), it seems likely that stem flow is high. Water appears to be channelled by the architecture of the leaf surfaces and flows towards the centre of the rooting zone.

Leaf hairs may also have a special role in accumulating occult water ephemerally present as dew or mist. A hair or pointed leaf lobe or tip will cool to dewpoint before the leaf surface. Hence droplets of occult water will build on these points. The question is whether such water flows to the leaf surface for cuticular absorption or merely re-evaporates when the energy balance alters. Both these phenomena require further investigation.

Figure 3.4: Density of *Larrea divaricata* (creosote bush) plants from twelve sites in California as a function of site rainfall. Redrawn from Woodell, Mooney and Hill (1969)

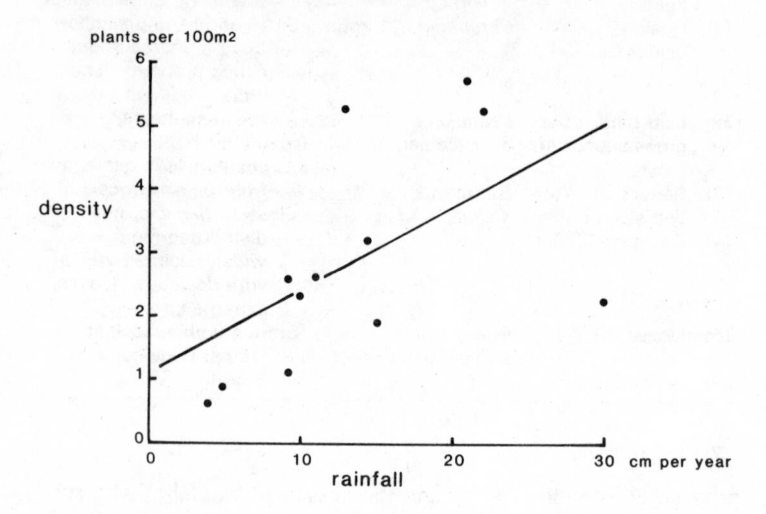

CONTROL OF WATER LOSS

There are two particular penalties to reduction of water loss, reduction of photosynthesis and increase in heat load. The latter arises from reduced evaporative cooling. Inspection of plant surfaces reveals adaptations that minimise water loss and reduce heat load. Table 3.2 summarises these adaptations.

DESICCATION TOLERANCE

Leaves are the organs which must bear the brunt of desiccation, and few leaves survive loss of more than 30% of leaf water. When this occurs in Chilean sclerophyll shrubs, leaf water potentials fall to less than −5.0 MPa (Dunn, Shropshire, Sung and Mooney 1976). In this case, recovery of photosynthetic activity occurs on restoring soil water supply.

There are a few records of recovery from greater degrees of absolute desiccation in mosses (e.g. *Tortula*), lichens (many types), ferns (e.g. *Ceterach officinalis*) and some angiosperms including *Ramonda nathaliae*.

The literature in this area speculatively tends to link the

Table 3.2: Leaf-surface adaptations to minimise water loss and reduce heat load

	Feature	Mechanism
(1)	Heavily cuticularised epidermis	Wax plates reduce diffusion of vapour from cell wall. Liquid water may be able to penetrate, e.g. rainfall or dew
(2)	Leaf hairs, stomata in pits or grooves, leaf rolling, rosette habit	Increase in boundary layer resistance by stabilising air layer at leaf surface
(3)	Reduction in stomatal number and size of stoma	Reduced cross-section for gas exchange
(4)	Deciduous habit	Leaf loss greatly reduces transpiration, but green stems may permit basal photosynthesis
(5)	Microphyllous habit	Small leaves are more readily cooled by convection
(6)	Reflective surfaces, viz. hairs, cuticle, epidermis	Heat load minimised with reflection of long-wave radiation
(7)	Sclerophylly	Long-lived leaves with heavy cell walls. Photosynthesis can occur during periods of high temperature. Must be accompanied by (a) deep roots, and (b) other water control features. Very characteristic of mediterranean-type climate. See Chapter 16

phenomena of water loss to the atmosphere, to extracellular ice and to osmotic systems. In all three cases the protective effects of metabolites such as sucrose, proline and other nitrogen- and sulphur-containing compounds are hypothesised. Enzyme membrane and organelle protection through the chemical and colligative properties of these metabolites is the central theme of this hypothesis. This attractive idea, of a unified system of compatible protective metabolites, has a long distance to run before it can be believed with confidence. Testing this hypothesis will entail unusual linkages between biochemistry and ecologists interested in extreme environments. The best explored case is that of the halophytes, and this will be further examined in Chapter 18.

AVOIDANCE OF DESICCATION

Two habits are entailed, the annual habit (therophyte), and the cryptophyte habit where the plant is protected by a soil covering during the arid season. In both cases a key item is the cueing of growth of seeds, bulbs and tubers to the onset of a rainy period. This enables interpretation of a wide range of growth patterns observed in living plant collections, including the low-temperature requirements for flowering of many bulbs and succulents. Annual species may have a leachable germination inhibitor associated with the seed coat, ensuring that germination commences only when soil is completely wetted.

A WIDER VIEW OF ARID ENVIRONMENTS

Ecological situations requiring economy of water are not confined to the tropics, or to situations with low soil water storage. Arctic and alpine plants may experience situations of adverse water balance. A common situation is exposure of shoots to dry air when roots are frozen. In polar deserts and rain-shadow zones of montane areas, a totally conventional arid regime may occur, namely low precipitation with high potential evaporation. The forms of many arctic and alpine plants may reflect this situation. For example *Dryas octopetala,* a widely distributed species in the northern circumpolar region, is typically found where winter snow cover is slight and soil is shallow. Its evergreen leaves are tough and glossy and have undersurfaces covered with densely felted hairs. It is a long-lived species with a woody rootstock exploring a large soil volume. Its water relationships deserve investigation in a wide range of altitudes and latitudes. In this and many other cases the investigation would need to be coupled with several aspects of mineral nutrition. Indirect effects of low soil moisture on nutrient turnover and soil microbiological processes will continue to be of interest.

4

Symbiotic and other associations for nutrient capture

EXPECTATIONS

Symbiosis between green plants and microorganisms is now known to be so widespread that the exceptions among green plants should be regarded as special cases. The basic unit of study must be conceived of as being a pair or cluster of organisms. A commonplace example is of a nodulated legume with a vesicular–arbuscular mycorrhiza. This adds a new dimension to plant physiology, agriculture and ecology. Without even considering the joint metabolic properties of symbiotic associations, we must realise that soil environmental conditions surround and influence all parts of the association. This may mean, for example, that anaerobic soil conditions may incapacitate or eliminate mycorrhizal fungi, which are obligate aerobes, more readily than a larger green plant possessing aerenchyma. Awareness of such possibilities should influence the formulation and testing of hypotheses in ecology more than it does at present. Detailed possibilities will be referred to under appropriate sub-headings. There has been a large amount of research activity and review in the area of plant–microbe symbiosis. This has concentrated on the fundamental biology of interactions and their economic importance. The ecological literature is relatively sparse. The account that follows will outline the expectations of the ecologist from each category of symbiotic interactions, referring to purely ecological material when it is available.

There are many avenues of study associated with these symbiotic relationships, biochemical, systemic, ultrastructural and ecological. In this account, attention will be focused on (a) the effect of the symbiosis on higher plant mineral nutrition; and (b) the effect of soil conditions on the relationship.

Table 4.1: Mycorrhizal systems (nomenclature and features after Smith 1980)

Nomenclature and features	Higher plant groups	Fungus	Ecological notes
Vesicular–arbuscular (VA); No sheath; extramatrical hyphae; intracellular 'vesicles' and 'arbuscules'; lysis	Most herbaceous families; ferns; many woody plants including tropical and temperate forest, angiosperms and gymnosperms, e.g. Myrtaceae, Magnoliaceae, Mimosaceae, Oleaceae, Rubiaceae, Verbenaceae, Euphorbiaceae, Cornaceae, Auraucariaceae, Cupressaceae, Taxodiaceae	Phycomycete: Endogonaceae	The symbioses between the Endogonaceae and most groups of vascular plants is of overwhelming importance. These symbioses are in an early stage of ecological appreciation, and potential commercial applications are relatively undeveloped. See Saunders, Mosse and Tinker (1975)
Ectotrophic: sheath, Hartig net No intracellular penetration or lysis	Trees: Pinaceae; Fagaceae–*Quercus, Fagus* Myrtaceae–*Eucalyptus*	Basidiomycetes and Ascomycetes; Phycomycete: Endogonaceae	Of great importance in temperate and boreal forests. The report of a phycomycete-ectotroph associated with *Eucalyptus* (Warcup 1975) will doubtless be investigated further because of the ecological and commercial importance of this genus.

Ericoid: sheath rare, extramatrical hyphae, intracellular coils	Ericaceae, Epacridaceae, Empetraceae	'Dark sterile' mycelia, e.g. *Pezizella* and *Clavaria*	Of particular interest to ecologists concerned with these very specialised situations
Arbutoid: sheath, no Hartig net, extracellular haustoria	*Arbutus* and *Arctostaphylos*; heterotrophic types, e.g. *Monotropa, Yoania*	Basidiomycetes, Ascomycetes. Fungi shared with ectomycotrophic hosts, e.g. forest trees	
Orchidaceous: external hyphae, intracellular coils, lysis	Autotrophic and heterotrophic Orchidaceae	*Rhizoctonia* spp. which are also strong saprophytes	The sheer size and diversity of the Orchidaceae gives this system ecological and commercial importance
Weakly mycorrhizal groups: few species only reported: most are negative when examined	Cyperaceae, Juncaceae, Proteaceae, Polygonaceae, Cruciferae	Where infection is reported, it is of the VA type	There have been few attempts to investigate weakly mycorrhizal types systematically (Baylis 1975) but their ecological importance justifies such an approach

MYCORRHIZAS

Mycorrhizal systems are briefly reviewed in Table 4.1. This focuses attention on the endotrophic types as being the most ubiquitous and yet needing most attention. Ectotrophic types may prove in the end to be a relatively specialised configuration, which benefits forest trees with long-lived roots. The morphology of the roots of the Ericales is somewhat unusual, being very slender and fragile. They ramify through soil, rather than thrust. Their mycorrhizas are thus different from those of other types, seeming to share the slender hyphae of the ectotrophic organisms with a habit similar to that of endotrophic systems. The chlorophyll-less heterotrophic types are directly connected by common mycorrhizal hyphae to forest tree 'hosts'. In the case of the orchids, their symbiont is a saprophyte, and probably provides carbohydrate to the developing seedling. The weakly mycorrhizal types need ecological evaluation. At present there seems no coherent ecological pattern, with both strongly oligotrophic groups, e.g. Cyperaceae, and ruderals, e.g. Polygonaceae, included.

Mycorrhizas and mineral nutrition

The predominant effect of all types of mycorrhiza is to improve the supply of phosphate to the higher plant by a factor of between five and thirty times. The main mechanisms for this increase are:

(a) Substantial increase in potential absorbing surface area, exploiting available phosphate in a larger soil volume (Table 4.2).

    Although mycorrhizal hyphae have a similar diameter to root hairs, their length may extend for some centimetres into the soil (see Figure 1.2). Furthermore, their active life is much longer. Most studies on the form of phosphate absorbed by mycorrhizas show that the same pool of inorganic phosphate available to non-mycorrhizal roots is being utilised (Tinker 1975).

(b) Phosphate storage occurs. A literature is developing that shows that storage of polyphosphate is a regular occurrence in mycorrhizal systems. The significance of this is probably in the assimilation of short-term flushes of available phosphate. These may not necessarily occur when conditions for utilisation and growth are optimal. This is called the 'overplus' situation

Table 4.2: Length and surface area of 1 g freshweight of root or mycorrhizal tissue assuming a density of 1 g cm^{-3}

	Diameter, mm	(μm)	Length m^{-1}	Surface area $cm^2\ g^{-1}$
Many unthickened roots	1	(1 000)	1.2	40
Fine roots	0.5	(500)	5.1	80
	0.25	(250)	20.4	160
	0.1	(100)	127.4	400
Coarse mycorrhizal hyphae	0.025	(25)	2 040	1 600
Fungal hyphae and root hairs	0.01	(10)	12 740	4 002
Actinomycete hyphae	0.002 5	(2.5)	204 000	16 000 ($1.6\ m^2$)

described by Harold (1966) for polyphosphate synthesis by free-living microorganisms.

(c) Phosphatase activity. Extracellular phosphatases can hydrolyse organic phosphate esters. Phytin (inositol hexaphosphate salts) is the predominant ester in soil (see Chapter 10), but, because of its chemical stability, phosphate availability is very low. It is an attractive idea that this non-available phosphate source is rather specifically attacked by enzymes from mycorrhizal hyphae. The evidence for this is not complete. For example, ectomycorrhizal roots of introduced *Pinus radiata* were shown to hydrolyse phosphate esters but there was no evidence on utilisation (Bowen 1973). Roots of grasses taken from the field have been found to possess phosphatase activity, but the endomycorrhizal aspect was not investigated (Woolhouse 1969). Furthermore, in neither case were surface microbial populations considered.

Rather specific investigations are needed to integrate the substrate, symbiotic and assimilation aspects of the utilisation of organic soil phosphate. This is an important area, first because it is to some extent in conflict with the finding in (a) above. Secondly, as soil phosphate concentration falls, there is a tendency for the proportion of organic phosphate to rise. Thus, in the most oligotrophic substrates, organic phosphate may represent 50% of the total phosphate resource.

Nitrogen assimilation

There is substantial evidence that mycorrhizas function effectively in the absorption of nitrate and ammonium from their substrate (Smith 1980). It has been suggested that this is of critical importance when ammonium is the main ion available in acid heathland communities, e.g. Stribley and Read (1975), and in woodlands where nitrification is suppressed. Plants with nitrogen-fixing root nodules of the *Frankia* or *Rhizobium* types are also mycorrhizal: endotrophic, ectotrophic or even of both types. The web of relationships, in cause-and-effect terms, between supply of nitrogen and phosphate, and growth of the autotrophic higher plants and the symbionts may prove to be of great ecological complexity.

The most simple case is to assume that limitation on growth and energy assimilation by phosphate supply is relieved by mycorrhizal infection. Photosynthate is then available to support the growth of nodules and for fuelling subsequent nitrogen fixation. Alternatively it may also be imagined that, for a phosphate-rich legume seedling, nodule formation and nitrogen fixation are a first priority. Mycorrhizal development may then follow, consequent on flow of photosynthate and organic nitrogen.

It is also interesting to note that the non-nodulating tropical forest members of the Caesalpinoideae, where investigated, are nearly always strongly ectomycorrhizal. Supply of mineralised nitrogen through a mycorrhizal system is clearly of importance in this group, and should not be underestimated in others (Corby, Pohill and Sprent 1983).

NITROGEN-FIXING SYMBIOSES AND OTHER ASSOCIATIONS SUPPLYING NITROGEN TO HIGHER PLANTS

The ecological advantage of a nitrogen-fixing symbiosis is the close coupling of the fixed nitrogen input to the metabolism of the higher plant, eliminating the nitrogen mineralisation step. The transfer of energy from photosynthesis to fixation systems is also very direct, and eliminates competition from non-fixing heterotrophs. In Table 4.3 the known distribution of root-nodule-forming symbiotic associations in vascular plant taxa is set out. The presence of nodule structures is not evidence, in the most rigorous sense, of nitrogen-fixing ability, but positive tests for nitrogen fixation have been achieved for most nodulated genera, if not for all species. The search for nodule systems is far from complete, and the taxa listed indicate the scope of studies so far.

Nodulated species are of special ecological importance in situations where soil nitrogen supply is low for a variety of reasons. Situations of special importance are:

— newly developing soils, e.g. sand dunes on the shores of oceans or lakes, volcanic lavas or ash beds, substrates exposed by glacial retreat, land slips, soil erosion or braided river beds;
— intrinsically low-fertility substrates, e.g. peatlands, leached mineral soils;
— ecosystems with high nitrogen turnover or removal, e.g. tropical rainforest;
— fire-dominated ecosystems, e.g. chaparral;
— where man's influence has produced: intensively cropped agroecosystems, and landscapes devoid of soil nitrogen because of construction, extractive industry or deforestation.

The statistics and ecological notes set out in Table 4.3 do not do justice to the wealth of biological detail concerning the morphology, microbiology, biochemistry and serology of nodulation. That a viable nitrogen-fixing nodule arises out of an infection of a higher plant by a soil microbe must be considered a remarkable happening in evolutionary terms. Bearing in mind the frequent role of nitrogen insufficiency in ecology, however, it seems curious how spasmodically nodule formation is distributed through plant taxa.

Apart from *Datisca* and the herbaceous legumes, herbs and monocotyledons do not appear to possess nodule systems. The nodulated woody species thus stand out as very remarkable plant groups that have overcome one extreme soil condition, lack of fixed nitrogen. It is worth commenting on each group in turn.

Cycad type

All the cyads bear nodules on soil surface roots which enclose colonies of photosynthetic cyanobacteria. An obvious speculation is that this form of symbiosis would have had greater ecological importance in pre-Cretaceous periods when this group had greater dominance.

Parasponia type

Nodules were noted in specimens of *Parasponia parvifolia*, a small tree in the Ulmaceae, collected in 1934. Nearly 40 years later it was demonstrated that the nodule microsymbiont was a slow-growing *Rhizobium* strain. *In situ* nitrogen fixation was shown subsequently to be similar to the perennial nodules of non-legumes. *Parasponia*

Table 4.3: Distribution of nitrogen-fixing root nodule systems in vascular plant taxa

Higher plant group					Examined		Microorganism	Ecological notes
Nomenclature	Order	Family	Genera	Total spp.	+ve	−ve		
Cycad type[a]	Cycadales	Cycadaceae	9	108	All	0	Cyanobacteria — *Anabaena* or *Nostoc*	Tropics and subtropics, of relatively minor importance in contemporary ecology, with *Macrozamia* listed as a pioneer species in Australia
Parasponia type[a]	Urticales	Ulmaceae	*Parasponia*	5	3	0	Eubacteriales: *Rhizobium* spp.	Pioneer species, e.g. on volcanic ash, in upland areas of Malaysia, Indonesia and Papua New Guinea
Alnus type[a]	Casuarinales	Casuarinaceae	*Casuarina*	48	18	0	Actinomycetales: *Frankia* spp.	Pioneer or coastal tree species in SE Asia and Australia, dwarf species in heath
	Coriariales	Coriariaceae	*Coriaria*	15	13	0		Temperate and montane scrub species
	Fagales	Betulaceae	*Alnus*	35	33	0		Arctic–alpine pioneer species and temperate wetland trees
	Cucurbitales	Datiscaceae	*Datisca*	2	2	0		Herbaceous species; disjunct distributions, Himalayas and California; subalpine stream beds, sand dunes
	Myricales	Myricaceae	*Myrica*	35	20	0		Bog species tolerating low pH; pioneer species in various situations
			Comptonia	1	1	0		

	Rosales	Rosaceae	*Rubus*	> 200	1	3		*Rubus ellipticus* from Java and Indonesia is the only nodulated species
			Dryas	4	3	0		Arctic-alpine pioneer
			Purshia	2	2	0		Chaparral shrub
			Cercocarpus	20	3	0		Chaparral, subalpine and desert shrubs
	Rhamnales	Eleagnaceae	*Eleagnus*	45	14	0		N temperate pioneer species
		Rhamnaceae	*Hippophaë*	3	1	0		Coastal shrubs
			Shepherdia	3	2	0		Desert and subalpine shrubs, arctic–alpine pioneer
			Ceanothus	55	31	0		Chaparral and desert shrub
		Colletiaceae	*Trevoa*	3	1	0		South American scrub species
			Discaria	10	5	2		
			Colletia	17	3	0		
Legume type[b]	Leguminales	Leguminaceae 640 (Caesalpin-oideae		17 000 30% nodulated			Eubacteriales: *Rhizobium* spp.	Tropical (rainforest species poorly nodulated) trees and shrubs
		(Mimosoideae)		96% nodulated				Xerophyte shrubs, temperate rainforest trees
		(Papilionoideae)		98% nodulated				Some xeromorphic shrubs, e.g. *Ulex, Cytisus,* many diverse herbs – coastal sands, grasslands, subalpine pioneer species, e.g. lupins. Very great agricultural importance

[a] After Akkermans and van Dijk (1981).
[b] After Corby *et al.* (1983).

thus joins *Casuarina* and leguminous trees as species of ecological and practical importance in reforestation in the tropics. Undoubtedly the search for other nodule-bearing tropical members of the Ulmaceae will continue, although it has not been fruitful so far (Akkermans and van Dijk 1981).

Alnus type

The taxa listed fall into two groups:

(a) Small families in which all species in a genus bear nodules; Casuarinaceae (one genus), Coriariaceae (one genus), Eleagnaceae (three genera) and Myricaceae (two genera). Here it may be argued that there is a coherent evolutionary pattern and the ecological distribution is also rather narrow.
(b) Families where nodulation is restricted to certain genera or species. These groups all present enigmas. For example, in the 58 genera of the Rhamnaceae, four genera are nodule bearing. Did other genera have nodules in the past, and do other genera have a potential for symbiosis if exposed to an appropriate endophyte? Are ecological niches available which may encourge nodulation in other genera?

In the genus *Dryas* (Rosaceae) at least three species bear nodules and are of ecological importance as pioneer species. However, *Dryas octopetala* in Europe and Japan does not bear nodules. Is this for biological or ecological reasons, or for some reason concerned with a separation of the species from its endophyte during the Pleistocene?

Although relatively few genera possess *Alnus*-type nodules compared with the legumes, all seem to be species of key ecological importance. Being woody, long-lived and widely distributed, they appear to occupy niches not exploited by legumes, in particular those in which another extreme environmental factor is additional to low nitrogen supply, e.g. waterlogging, drought, low pH, other major nutrient deficiency, or a subalpine or tundra climate.

Legume type

The sheer size of the legume group indicates how strongly the evolutionary current leading to symbiotic nodules has flowed. Within the legume family the three subfamilies appear to show a trend from a relatively primitive tropical group with a low frequency of nodulation, Caesalpinoideae, to adaptive radiation into a wide range of situations. The Mimosoideae is a mainly woody group of special importance in the southern hemisphere. The Papilionoideae, although with some woody genera, contains some thousands of herb species. Many are of great economic importance as the peas, beans, pulses and peanuts of commerce.

ALTERNATIVE NITROGEN-FIXING ASSOCIATIONS

A series of associations between nitrogen-fixing heterotrophic bacteria and plant surfaces is known to exist. Secretions from the plant surface are thought of as augmenting the carbon supply to the heterotroph. Secretion of fixed nitrogen is then leaked as an extracellular product to the plant surface. It is estimated that this contribution to the plant community is quite small, in the order of a few kilograms of nitrogen per hectare per year, but may be significant if nitrogen is limiting plant production (Giller and Day 1985). It is not quite clear if mineralisation is always necessary before absorption of fixation products can occur.

The rhizosphere of dune grasses is a striking example where, in the case of *Ammophila arenaria* in Britain and New Zealand, nitrogen fixation is associated with root surfaces. Enriched populations of *Azotobacter* appear to be the answer. It remains to be seen how widespread this phenomenon is, but records also exist for *Oryzopsis*, a desert grass, and *Elymus junceiforme,* a salt-tolerant fore-dune grass. In the latter case the nitrogen-fixing organism is *Bacillus polymyxa*. Several tropical grasses have been investigated in an agronomic context (Giller and Day 1985). The role of mycorrhizal hyphae in absorbing nitrogen may be important, as it is known that organic nitrogen can serve as a carbon source. Ammonia would be released into the rhizosphere as a consequence. Root surface nitrogen fixation is also well documented from forest soils. In fact, a record of nitrogen fixation by 'root nodules' in the conifer *Podocarpus* is now accepted as being due to surface heterotroph populations.

Aerial surfaces also bear populations of nitrogen-fixing bacteria. Whereas in temperate situations their activities are slow and their presence is virtually accidental, in tropical vegetation the situation may be very different. In a humid environment parts of the plant architecture encouraging accumulation of leaf washings or stem flow may well be found to be enriched with nitrogen-fixing micro-epiphytes. Important associations exist between higher plants and cyanobacteria. The aquatic fern *Azolla* bears colonies of *Anabaena azollae* in the dorsal lobes of its leaves. This association is widely distributed in tropical and subtropical zones and is of agricultural importance as a green manure.

In concluding this section the agronomic importance of bacteria and cyanobacteria in rice cultivation must be briefly mentioned. Although this is a system managed for a long period of human history, it must be considered to be part of the ecology of rice-growing regions. Studies of rice culture also provide a model for tropical, semi-aquatic ecosystems (see Watanabe and Brotonegoro 1981).

Rice seedlings are planted into flooded soils, and four functional ecological sub-units rapidly differentiate:

(a) Flood water. A mixed community of algae, cyanobacteria, weeds and heterotrophs. Some nitrogen fixation occurs. Marked diurnal fluctuations in aeration as photosynthetic oxygen is produced and consumed.
(b) Soil surface. Redox generally greater than +300 mV in a layer 2–20 mm thick. Aerobic bacteria predominate, especially nitrifiers.
(c) Subsurface soil. Anaerobic conditions predominate, with activities linked to the intensity of reducing conditions (see Chapter 9). Nitrogen fixing *Clostridium* spp. may be very common.
(d) Rice plant and its rhizosphere. The rhizosphere is aerobic, with the rice plant acting as a channel for gas exchange. A range of nitrogen-fixing heterotrophs have been isolated from the root surface including *Azotobacter*, *Spirillum* and *Clostridium*.

TESTING FOR NITROGEN FIXATION

A number of approaches exist for deciding if nitrogen fixation is occurring in a natural system. These are now used in two major ways. First, the search continues for new nitrogen-fixing associations, the unit of study being the individual plant or organ. Secondly, there is a need to make quantitative estimates of nitrogen input into natural and managed ecosystems.

(1) Testing for sustained growth on nitrogen-free media. This approach is obviously most suitable for testing microorganisms

in closed cultures. As a rough screening test for higher-plant associations it still has merit if users are aware of nitrogen inputs from rainfall, dust, insects, etc.

(2) Measurement of gain in total nitrogen content by the unit of study. This approach is of particular value in studying pioneer vegetation. In steady-state situations this rate of gain may be offset by losses not readily detected.

(3) Incorporation of ^{15}N-labelled dinitrogen by the system. This is the most rigorous test for establishing the reality of nitrogen fixation. It is especially recommended as a critical test where a novel biological system is suspected of active dinitrogen fixation. In this test, the system is exposed to ^{15}N-labelled nitrogen gas in air for an appropriate time period. The system is then examined for the presence of ^{15}N-labelled organic nitrogen compounds.

This examination requires standard biochemical extraction and concentration procedures followed by ^{15}N determination. As ^{15}N is a stable isotope it must be detected in the face of a large excess of stable ^{14}N using mass spectometry. The main barrier to widespread use of this procedure is its relatively high cost. The capital cost of equipment is high and the unit cost of ^{15}N is also substantial. Because the procedure is cumbersome, the rate of experimental progress is slow. The ^{15}N test is not especially sensitive and may not be suitable for slow systems or small units.

(4) The acetylene reduction test. At the heart of all biological nitrogen fixation systems are the so-called nitrogenase enzymes which catalyse the reduction $N_2 \rightarrow 2NH_2$. The gas acetylene is also reduced to ethylene by the nitrogenase enzymes, i.e. $C_2H_2 \rightarrow C_2H_4$. This phenomenon was utilised to produce a practical assay system by Hardy, Holsten, Jackson and Burns (1968). It has now been used with all known biological nitrogen fixation systems.

An active nitrogen reducing system will reduce acetylene at a rate proportional to the nitrogen assimilation rate. The gases acetylene and ethylene may be easily and rapidly estimated using gas–liquid chromatography (GLC) with a flame ionisation detector. The conversion ratio of acetylene to be reduced to nitrogen fixed is of practical interest. This ratio depends on the electron requirements for the two reactions, two for acetylene and six for nitrogen. Thus the observed rate of acetylene reduction has to be divided by three (see Figure 4.1). The literature

Figure 4.1: Practical steps in determining acetylene reduction rate of nitrogen fixing systems (see Table 4.4 for details)

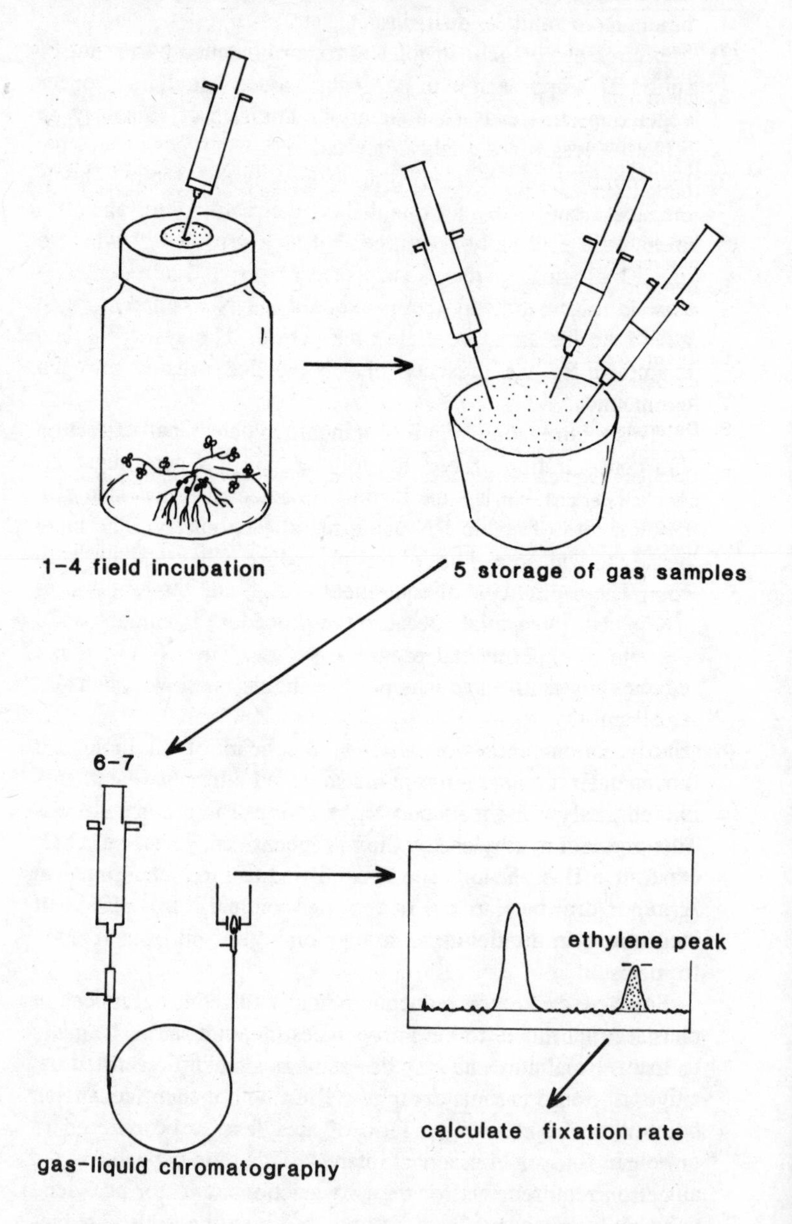

Table 4.4: Using acetylene reduction to determine nitrogen fixation rate in the field (see Figure 4.1)

1. Select unit of study, e.g. a small leguminous plant.
2. Place in incubation chamber. A glass jar with a membrane of tyre rubber fitted to the lid is convenient.
3. Inject 5 ml of acetylene. (This may be taken into the field conveniently in a small motor tyre inflated from a cylinder of the gas. Other workers have generated it in the field using the CaC_2 reaction with water.) The incubation chamber is fitted with a heavy membrane made from tyre rubber. Standard disposable hypodermic syringes are used throughout for gas handling.
4. Incubate *in situ* at field temperature. A period between 1 and 24 h is convenient.
5. Remove a subsample of the gas with a 1 ml syringe. Store sample by inserting needle of filled syringe into a rubber stopper. Transport samples to laboratory.
6. Inject gas sample into gas–liquid chromatograph, having first calibrated the system with standard gases.
7. Record ethylene peak and calculate concentration.
8. Determine dry weight and any other required characteristics of test system, e.g. total nitrogen, water content, chlorophyll.
9. Calculate fixation rate on a suitable basis: milligrams of nitrogen fixed per plant per day = millimoles C_2H_4 per plant per day × 28 (molecular weight of N_2)/conversion factor of 3.
10. Calculate on a suitable field basis, e.g. mg N m^{-2} day^{-1}, i.e. multiply the above value by plants per square metre.

suggests slightly different ratios for different systems, and it is suggested that new systems should be calibrated with ^{15}N (Hardy, Burns, Hebert, Holsten and Jackson 1971). Another fundamental problem is the production of ethylene by plants and microorganisms in the absence of acetylene or nitrogen fixation. This problem is handled by using acetylene-free controls in the experimental design and subtracting indigenous ethylene values. A minor drawback is the need to use a sealed container for incubation. In the field this may give rise to problems arising from disturbance or overheating.

Some practical details are outlined in Table 4.4, but it is emphasised that this technique is very flexible and can be adapted to many field situations. The method is also sufficiently sensitive to be used in short experiments lasting no more than a few hours. Small units such as a few detached nodules or a fragment of lichen thallus may also be utilised. Flexibility and sensitivity allied to speed and relative cheapness make this technique very valuable to ecology.

CARNIVOROUS PLANTS

Introduction

To end this review of associations directly enhancing the mineral nutrition of higher plants, carnivory will be briefly mentioned. This is an evolutionary backwater, but full of interest and reinforcing the idea that nutrient relationships and plant evolution are tightly linked.

Carnivorous plants are usefully defined as species that can benefit from nutrient absorption from dead animals which are actively attracted, captured and possibly digested. Prey are typically small arthropods, but frequently include small molluscs and even vertebrates, e.g. tree frogs. The emphasis on active attraction and capture is required to avoid confusion with the passive benefit most plants enjoy from mineralisation of dead animals. This emphasis also distinguishes between defences against herbivores which may result in animals being immobilised or killed without positive gain.

The carnivorous plants may be regarded as a small ecological group for which the evolutionary imperative has been directed towards mineral nutrient gain from sources other than soil. In all groups the structures responsible for attraction, for capture and for absorption of ions are modified leaves. The armoury of elaboration includes the following devices:

(a) nectar-secreting glands, e.g. Sarraceniaceae, Nepenthaceae;
(b) backwardly directed hairs, e.g. *Sarracenia, Genlisea*;
(c) shiny surfaces, e.g. Sarraceniaceae, *Brocchinia*;
(d) adhesives, e.g. Droseraceae, *Pinguicula, Triphycophyllum*;
(e) rapid action, trigger-stimulated traps, e.g. *Dionaea, Aldrovandra, Utricularia, Biovularia, Polypompholix*;
(f) distinctive coloration, light transmission, reflection or refraction: most groups except aquatics.

These mechanisms may be considered as elaborate in their own way as the renowned pollination mechanisms of, for example, the Orchidaceae. It is not surprising that both attracted the attention of Charles Darwin. Plant carnivory is always evident on nutrient-poor substrates, including bogs, Australian heathlands and the epiphyte habitat.

Plant groups

The summary of carnivorous plant groups (Table 4.5) gives a picture of a relatively few plant groups, all advanced angiosperms as determined by flower structure, and surviving in unshaded nutrient-poor habitats. Where they occur in plant communities, several groups may be present. For example, in the Pocosin bogs of the Carolinas, USA, the endemic *Dionaea muscipula* is accompanied by *Pinguicula, Sarracenia* and *Drosera* spp; the carnivorous bromeliad *Brocchinia reducta* is accompanied by *Heliamphora, Drosera* spp., *Utricularia* spp. and *Genlisea.*

For all carnivorous plants studied, a rich natural history has been forthcoming. However, the nutrient studies concerning any are far from complete. We do not clearly understand the reward, in terms of nutrient budgets, of this evolutionary effort. This is in spite of physiological studies showing enzyme secretion and ion absorption by the leaves of some groups. These hint at their ecological potential. The ecological question to be addressed is 'Does ion absorption from prey account for a high proportion, i.e. greater than 90%, of the ion budget for the lifetime of the plant?' It would be of interest to determine if this mass balance altered systematically from group to group.

Another problem is the blurred distinction as to mode of mineral nutrition that applies to the bromeliads and possibly other tropical epiphytes. Between the review of Benzing (1973) and that of Girnish, Burkhardt, Happel and Weintraub (1984), there is evidence that the leaf-base tanks of this group have many roles, e.g. collecting debris; attracting chironomid species, which pupate in the tank, or ant colonies; supporting microcosms of saprophytes; and attracting and trapping other insects which eventually decompose. Ions released in the course of any of these activities may be absorbed by leaf bases, or leak out to be absorbed by roots.

There may be several opportunities for import of ions into the bromeliad system as rainfall, dust, debris or insects. Mineralisation could be encouraged by microbes or other fauna. Even nitrogen fixation by associated free-living microbes is a possibility that should be examined in tank bromeliads and the pitcher plants.

The aesthetic attractions of carnivorous plants, illustrated well by Slack (1979), make them items of commerce frequently taken from the wild. Some species are threatened by such collecting, e.g. *Dionaea muscipula*, and require legislative protection. This species is classified as vulnerable in the *International Union for the Conservation of Nature Plant Red Data Book* (Lucas and Synge 1978).

Table 4.5: Carnivorous plant groups (after Slack 1979; Girnish *et al*. 1984)

Order	Family	Genera (spp.)	Trap type	Notes
Sarraceniales	Sarraceniaceae	*Heliamphora* (6)	Terrestrial pitchers	South America. Wet sites in tropical–subtropical uplands.
		Sarracenia (9)	Terrestrial pitchers	North America. *S. purpurea* naturalised in Ireland. A wide variety of pitfall traps
		Darlingtonia (1)	Terrestrial pitchers	Endemic, coastal and mountain bogs of Oregon and N. California. No enzymic secretions
	Cephalotaceae	*Cephalotus* (1)	Terrestrial pitchers	Endemic, south-west and western Australia. Non-carnivorous foliage leaves and pitcher leaves
	Nepenthaceae	*Nepenthes* (71)	Suspended pitchers	Old World tropics except mainland Africa. Typically rainforest climbers or scrambling plants of drier scrub. Enzymes reported from some species
	Dioncophyllaceae[a]	*Triphycophyllum* (1)	Adhesive leaf appendages	This genus is a liane confined to African rainforest but widely distributed therein. Enzyme activity from glands is reported
	Byblidiaceae[a]	*Byblis* (2)	Adhesive hairs, passive leaves	Western Australia, Northern Territory, Queensland. Rootstock survives bush fires. Enzyme activity reported
	Droseraceae	*Drosophyllum* (1)	Adhesive hairs, passive leaves	Spain, Portugal, Morocco. Long-lived perennial. Semi-woody
		Drosera (90)	Adhesive hairs, reactive leaves	Large range of form and habitat — alpine, tropics; geophytes and scrambling. Worldwide distribution. Leaves form a cup when prey is caught. Enzyme secretion copious

		Dionaea (1)	Spring trap	Endemic to coastal bogs of North and South Carolina. Stimulation of trigger hairs by large insects causes halves of leaf to fold inwards. Bars at leaf edge prevent escape
		Aldrovandra (1)	Spring trap	Rootless floating aquatic with rosettes of traps like small *Dionaea* leaves. Widely distributed southern Europe, Africa, Asia, Australia
Scrophulariales	Lentibulariaceae	*Genlisea* (16)	'Eel trap'	Tropical, South America, West Indies, Africa, Madagascar. An aquatic group trapping by a hair-lined tunnel into a bulbous cavity in the petiole of modified leaves. Not well studied
		Pinguicula (48)	Adhesive glands	A group always on wet soils, humic or mineral, acid and base rich. Arctic, temperate and tropical. Northern hemisphere and Central America. Leaf glands well studied for enzyme secretion
		Utricularia (280)	Trigger-operated suction trap	Wide habitat range from peat-bog pools to bromeliad tanks
		Biovularia (1)	Trigger-operated suction trap	
		Polypompholyx (2)	Trigger-operated suction trap	Western Australia, South Australia, Victoria
Bromeliales	Bromeliaceae	*Brocchinia* (1)	Pitfall trap	A convincing detailed case is made for including this leaf tank bromeliad among carnivorous plants. Wet savannah and bogs, highlands of S Venezuela

[a] The systematic status of these two families is disputed. Cronquist (1981) assigns the Byblidiaceae to the Rosales, close to Pittosporaceae, and the Dioncophyllaceae to the Violales.

5

Herbivores, decomposers and other soil organisms

FEATURES IN COMMON

There are large differences between herbivores, decomposers and other soil organisms, and so their biologies are conventionally treated separately. When considering the movement of ions in the soil–plant system, however, there are features in common which are usefully discussed together. They may be linked by a simple food web as shown in Figure 5.1

(a) *Net mineralisation.* In any consumption of organic material by an organism, there are two main fates for the elements contained in it: either to be assimilated into the tissue or to be catabolised and eliminated as ions. A third fate is the elimination of residual organic matter in modified form. Thus all secondary consumption activities tend to release ions into the soil system.

(b) *The effect of energy : mineral nutrient ratios.* Heterotrophic organisms have characteristic energy, protein and mineral metabolite requirements. In herbivores this may lead to discriminatory feeding, metabolic or elimination patterns. For example, feeding aphids may remove and assimilate a higher proportion of protein from a predated plant than of carbohydrate, the carbohydrate-rich 'honey-dew' being eliminated to other heterotrophs. Similarly, the catabolic activity of heterotrophic fungi and bacteria may be restricted if the C:N or C:P ratio is too high. In a reciprocal manner, a decomposition system with a high C:N ratio may immobilise available nitrogen by acting as a sink. Ionic nitrogen would then be diverted to microbial tissue. Absolute low nutrient concentrations, found in bog or heath formations, for example, may prove to be an

Figure 5.1: The most simple association of herbivores, decomposers and other soil organisms

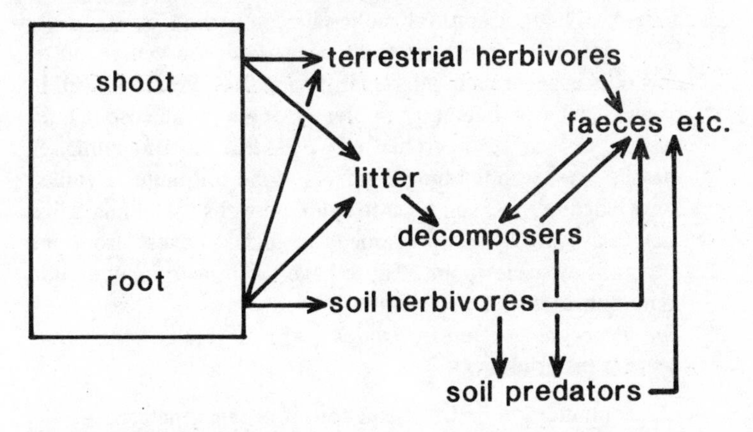

independent restricting factor in all decomposition processes. Organic matter will then accumulate as a direct result of nutrient deficiency.

(c) *Heterotrophic processes may occur in an episodic manner*. The reason for drawing attention to this apparently commonplace observation is that an episode presumably generates a pulse of nutrient availability. Examples include:
- — decomposer activity responding to soil temperature rise in temperate zone spring period, giving rise to a pulse of phosphorus availability (Tate 1984)
- — microbial population crash with soil drying (see Chapter 17),
- — transient insect herbivore attack
- — pollinator activity during the flowering season
- — grazing intensity linked to herbivore population size (see Chapter 12)
- — surface casting of earthworms (Tate 1984)

At the present time clear and quantitative links between all the above processes and mineralisation have yet to be studied in detail. However, there seems to be a rich area for exploration by workers interested in an interdisciplinary field. The work of Anderson, Huish, Ineson, Leonard and Splatt (1985), which examines the mineralising effect of soil animals on an oakwood soil, is discussed later.

(d) *The stepwise transformation towards mineralisation.* A well known example is the production of dung by a large herbivore and the subsequent fungal succession with disappearance of substrate energy fractions. The ultimate end of this process is the production of recalcitrant organic fractions which resist activity of heterotrophs (phenolic-tanned protein, for example). Simultaneously, transformations occur with the generation of new chemical forms in microbial tissue. The fungal cell wall in particular is rich in chitin, a polymer of amino-glucose. Chitin may be resistant to microbial enzymes, but specific chitinases may be present in the fungal grazers of the soil fauna. Detailed knowledge of most soil heterotroph food webs is so limited that they may have to be regarded as 'black boxes' from the mineralisation viewpoint. These have an organic input and an ionic output into the soil liquid phase.

(e) *The processes of mineralisation*. The processes that affect organic material are:

(i) comminution or fragmentation of organic material;
(ii) catabolism or metabolic breakdown of material;
(iii) leaching or dissolution or transfer of materials to soil water.

These three basic processes apply equally to the chewing of a living leaf by a grasshopper (herbivory) or the breakdown of a fallen dead tree trunk by a whole guild of fungi, bacteria and arthropods. The leaching process is important, as it may alter the tendency for reabsorption by the decomposer organisms by removing ions to lower horizons. Losses to all potential absorbers may also be caused by leaching.

THE DECOMPOSERS

About 80% of soil heterotroph biomass comprises fungi and bacteria (Table 5.1). Immediate problems of division of function arise. How much of this biomass is closely linked with the activity of the higher plant, i.e. mycorrhizal hyphae and rhizosphere bacteria? The case has already been made for recognising this tight coupling of current photosynthesis with nutrient accumulation as a special case, different from other heterotrophs. For a clear answer the identification of fungal hyphae and bacterial cells in soil will need to progress considerably.

The next largest contributors to soil biomass are the earthworms (annelids) and the enchytraeids. Their food intake is relatively easily specified as comminuted organic matter plus its attendant microflora. Worm casts contain more available nitrogen and phosphorus

Table 5.1: Decomposers and other soil organisms with an approximation for biomass. Data from several sources

% Soil biomass	Organism group	% Decomposer biomass
66.0	Roots[a]	–
17.1	Bacteria	55.5
13.1	Fungi	38.9
1.4	Actinomycetes	4.2
0.1	Algae[a]	–
0.7	Protozoa	1.9
0.2	Nematodes	0.3
0.7	Earthworms and enchytraeids	2.0
0.3	Collembola, Acarina, Myriopoda, Opiliones, Hymenoptera, Diplopoda	1.0
0.3	Diptera, Crustacea	0.8
0.1	Vertebrates	0.4

[a] Primary producers.

than bulk soil. Furthermore, mineral soil from below the litter layer may be brought to the surface.

In some interesting direct experiments with earthworms and small soil arthropods, Anderson *et al.* (1985) have directly demonstrated mineralisation effects. In a particularly realistic lysimeter experiment, ammonium and nitrate were determined in leachate from containers with and without soil animals. The paired lysimeters contained woodland soil and litter, but, in the case quoted, no oak roots. Millipedes, woodlice and earthworms were added in April to one of each pair of lysimeters, equivalent to 14 g fresh weight per square metre. Both plus and minus animal treatments were assumed to possess similar microbial, micro- and mesofaunal populations. No differences in nitrogen release were recorded for the first 14 weeks of the experiment. However, in the period July to mid-November the cumulative release depicted in Figure 5.2 was achieved. In this period a 34% increase in mineralisation was seen to be associated with the presence of fauna. In both treatments a number of flushes of soluble nitrogen accompanied drying and subsequent wetting.

Experiments with microcosms and lysimeters have shown so far that the enhanced mineralisation is fairly simply related to biomass of animals, irrespective of taxonomic group. The activity of animals is, however, clearly complementary to that of microbes. 'Conditioning' of litter by fungi facilitates attack by fauna, probably another manifestation of a succession. Fauna also seem to be able to consume litter and liberate nitrogen at C:N ratios normally leading

Figure 5.2: Cumulative nitrogen losses from small lysimeters in an oak woodland, July to mid-November. Data from Anderson *et al.* (1985)

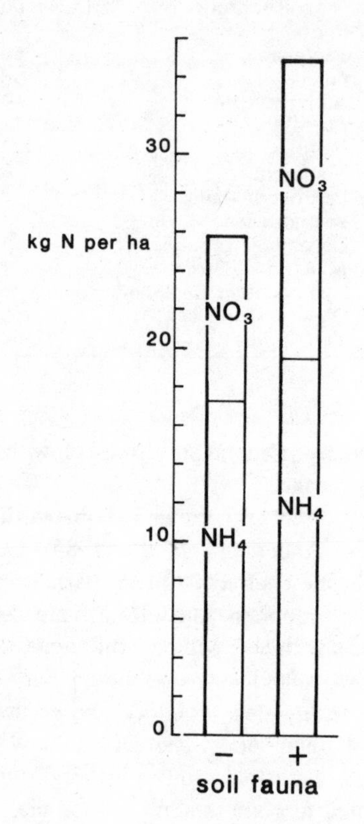

to immobilisation of nitrogen by microbes.

The activity of all soil microorganisms is strongly modified by soil temperature and water potential. Soil pH is another important variable. Soil fungi appear to resemble higher plants in that their respiration, a measure of cellular activity in culture, is not greatly affected by pH over the range pH 5–8 (Griffin 1972). In soils there is a marked trend towards a greater number of species and more colonies with higher soil acidity. Furthermore, fungal species show distinct distributions with soils of particular pH, again resembling higher plants. Evidence also exists that fungi can influence the pH of their milieu. A well-known example is the raising of pH through the release of ammonium ions when glucosamine, from the chitin of fungal hyphae, is decomposed by an actinomycete (Williams and

Mayfield 1971). This parallels the alteration of pH brought about by root activity in terms of cation absorption or carbon dioxide production.

Bacteria relate to soil pH in much the same way, with distribution of species and balance of metabolic activities being correlated with pH of bulk soil. In ecological studies of progressively smaller organisms it is less easy to distinguish between their microniches, or to perceive complex interactions between populations of organisms and between them and their surroundings. The role of sulphide-oxidising bacteria in acid production, and its profound ecological effects, are referred to in Chapter 14.

HERBIVORY AND THE SOIL–PLANT SYSTEM

To the plant biologist, herbivores are an exciting set of animals. They are diverse, specialised and seem to have coevolved with the plants they predate. The reason for presenting the long list in Table 5.2 is to expand appreciation of the pervasive nature of herbivory.

To the individual plant such predation can have a variety of consequences. These include stimulation of the plant through compensatory growth, prevention of reproduction by seed as a result of flower or seed predation, dissemination of seed by fruit predation, pollination through nectar or pollen predation, and infection by viral (e.g. tobacco mosaic virus) or fungal (e.g. Dutch elm disease) pathogens. Examples of more broadly ecological consequences are the effects on landscape of beavers or the vegetation destruction brought about by grazing of goats or elephants.

'How does grazing have any influence on the nutrient cycle?' is the main question to be addressed. By adopting a strategy of predating the base of the biomass-energy pyramid, herbivores have potential access to abundant energy as carbon components but a dilute source of minerals. As Mattson (1980) points out, whereas animals comprise about 7–14% by weight of nitrogen, plants are largely in the range 1–7% nitrogen. The other elements essential for animals are in an approximate correct balance, however. Occasional exceptions have been noted, for example the low Na:K ratios in various tundra species and the remarkably low calcium concentrations in *Eriophorum* (cotton grass) from various habitats (Chapin, Miller, Billings and Coyne, 1980). Concentrations of elements toxic to herbivores also occur from time to time, for example high selenium in the legume genus *Astragalus* which is referred to in Chapter 2.

Table 5.2: Examples to illustrate the diversity of herbivores

Mammals	Koala bear, a specialist feeder of *Eucalyptus* foliage
	Kangaroo and wallaby species, which are generalist grazers
	Most rodents are predominantly herbivorous
	Dugongs and manatees utilise aquatic vegetation
	The large and diverse ungulate group demonstrates the evolution of a four-chambered ruminant stomach
	The primates exhibit a wide range of feeding patterns, most being omnivores. However, many monkeys, the gorillas, chimpanzees, orang-utan and man are predominantly plant feeders
Birds	Many specialist seed and fruit feeders, e.g. parrots and finches
	Grazers, including some geese
	Humming birds are specialist nectar feeders
Insects	There is a great diversity of insect plant feeder types, which may be studied through the pest control literature in agriculture, horticulture and forestry. An interesting complication is the differences in feeding pattern between larval forms and adults. Because of mortality in the population, larvae are frequently the more important
	Modes of insect herbivory:
	nectar feeders: bees, butterflies and moths, flies, ants
	leaf chewers: moth and butterfly larvae, grasshoppers and locusts, adult beetles, e.g. Colorado beetle
	sap suckers: aphids, bugs, scale insects, leaf hoppers
	seed feeders: harvester ants, many beetles and especially weevils
	gall formers: larvae of flies, gall wasps, sawflies and moths
	tunnelling insects: bark beetles, stem borers, leaf miners
	root feeders: beetle larvae, root aphids
	wood chewers: termites

Herbivores of small body size overcome the problem of dilution of elements by very specialised feeding processes. Phloem sap is between 10 and 100 times richer in nitrogen than xylem, and, because of evaporation, nectar may be even better value (Mattson 1980). Buds, young leaves, fruits and seeds also have higher concentrations of mineral nutrients compared with carbon.

Such predation may have a debilitating effect on the plant concerned, and it is suggested that changing elemental ratios lead directly to the increased production of phenolic compounds in plants. Such compounds, which reduce the palatability of plant tissue, or its utilisation, are called 'allelochemics' (Rhoades and Cates 1976). The argument for a link between nutrient status of a plant and production of allelochemics is as follows. When carbon skeletons are being actively synthesised and utilised for growth, few phenolics are produced. When nitrogen or phosphate depletion leads to excess carbon compounds being synthesised, then carbon skeletons are available for phenolic production. A piece of direct physiological evidence exists, namely that sucrose added to nitrogen- and phosphate-starved *Acer pseudoplatanus* cell cultures, leads to phenolic synthesis. The addition of urea inhibits such synthesis (Phillips and Henshaw 1977).

Ecological evidence includes an interesting study of defoliated birch trees on a nutrient-poor soil. High levels of phenolics were produced for four years after defoliation. Phenolic production was reduced by experimental nitrogen addition. It was postulated that this induced allelochemic defence prevented repeated defoliation by moth larvae (Tuomi, Niemela, Houkioja, Siren and Neuvonen 1984).

Predation by large herbivores must follow a different pattern. They cannot take effective advantage of ephemeral flushes of high mineral content, except by timing reproductive events or migration. Neither can they graze in as highly selective a manner as smaller animals; their needs for energy are too great. Their strategy is to consume a high bulk of vegetation, digesting for a moderate recovery of energy and producing a high volume of faeces. Digestion is effective in releasing for assimilation in the gut the mineral nutrients contained in plants, especially nitrogen, phosphorus, potassium, calcium and sodium. Excretion of urine releases a proportion of these elements which varies according to the metabolic condition of the animal concerned. For example, a fully grown male mammal will return close to 100% of dietary phosphorus intake as urine, whereas a pregnant or lactating female or young growing animal will return much less. It is difficult to generalise on the direct nutrient return via urine and faeces but it will vary somewhat with element. Some estimates were made by Bulow-Olsen (1980) on nursing cows grazing on a marginal agricultural area in Denmark dominated by *Deschampsia caespitosa.* This was an unfertilised system, and thus the results are a useful yardstick for less managed situations. When compared as a percentage with the soil pool of 'available' nutrients on a square metre per annum basis, the returns by the animals represent: calcium, 0.9%; magnesium, 5.5%; potassium, 20.8%; and phosphate, 0.16%. Note that the domestic animals in this experiment are only one herbivore in the system. However, these observed values could well represent an important part of the turnover of the ecosystem and are of the same order of magnitude as losses to litter.

It is a common observation that the deposition patterns of urine and faeces are non-random. Ecological pattern may result from this behavioural phenomenon, especially from the accumulation of phosphate in the soil of deposition sites (see Chapters 13 and 17).

The large faeces of herbivores such as the African elephant and other savannah herbivores are effectively dispersed by dung beetles, possibly the most extreme form of a specialised invertebrate decomposer (Heinrich and Bartholomew 1979). The coprophilic fungal succession on dung is another manifestation that this source of carbon and mineral ions is rapidly exploited by specialised heterotrophs. Early colonisers such as *Pilobolus* disperse spores widely so that they are ingested by the grazing animal. Passage through the gut breaks spore dormancy, and germination of spores is followed by rapid growth and ultimate production of fruiting bodies.

These observations, combined with agricultural studies on animal manures and slurries produced by intensive stock raising (Glasser 1982), indicate that rapid mineralisation of ions in dung and urine occurs. We lack careful studies on the role of invertebrate herbivores in returning ions to the soil. This could be the prime pathway through which ions are shunted in very nutrient-poor ecosystems.

6

Vegetation and fire

Burning of vegetation must be considered to be a natural phenomenon in certain grasslands, shrublands and forest types. The influence of man has been to increase the frequency of burning of, for example, European bogs, the boreal forest and Australian heaths (Gill 1975). Man has also extended his range into ecological situations not subject to natural fires, e.g. by shifting cultivation in humid tropical forests. There are clearly many aspects to this topic in terms of the biological effects and their evolutionary significance (see Kozlowski and Ahlgren 1974; Wein and McLean 1983). However, there are a number of important effects on the soil–plant system operating at a nutritional level that merit discussion here.

PREREQUISITES FOR FIRE

1. Fuel

The probability of fires and their intensity are related to the quantity of burnable material. This is often non-living plant material which has accumulated against decomposition rate. Decomposition may be controlled by recurrent drought, lack of nutrients or presence of inhibitory substances. Low grazing activity can also encourage accumulation of dead biomass. With a high biomass a prolonged burning will lead to higher soil temperatures and more complete destruction of organic matter, dead and living. High biomass also leads to more extensive burned areas.

2. Flammability

Three factors influence the ease with which a natural source of ignition, such as a lightning strike, will start a fire. These are the water content, the temperature of potential fuel and the presence of volatiles with a low ignition temperature. Dry fuel is a feature of burnable grasslands whereas many shrublands and forests possess abundant resins and terpenoids.

3. Source of ignition

Many authors stress that man (*Homo sapiens*) may have employed fire as a food-gathering 'tool' since earliest evolutionary times. Tropical and subtropical grasslands in particular bear the pattern of game-management fires. Regular management burning of 'moors' in Europe is also well established in historic times. Lightning strikes are the main source of ignition in the absence of man.

4. The chemical and physical consequences of burning

These may be listed briefly as follows:

(a) The consumption of living and non-living biomass and soil organic matter. Degree of consumption is very variable, being dependent on biomass, flammability, and weather conditions such as temperature, wind speed and rainfall.
(b) Volatilisation of nitrogen and sulphur in biomass and soil. Between 25 and 60% of nitrogen may be lost.
(c) Effective mineralisation of phosphate, potassium, calcium and magnesium and of a portion of remaining nitrogen.
(d) Destruction of organic process inhibitors in soil, e.g. nitrification inhibitors, germination inhibitors.
(e) Maximising vulnerability to leaching of mineralised ions.
(f) Reduction in water-holding capacity and increasing runoff
(g) Losses of ash elements as particulates dispersed by convection currents.
(h) Because of changes in reflectivity, decreases in insulation and reduction of evaporative loss, there is a trend towards higher soil temperature.
(i) Decrease in slope stability and a tendency for accelerated particulate removal from watersheds.

5. Biological effects

These are as follows:

(a) Stimulation of growth of surviving species because of nutrient ion release and possibly through a range of indirect effects, e.g. predator control, decreased shade, higher soil temperatures, reduced competition. This is referred to as the 'ash-bed' effect.
(b) Germination of 'seed bank'. This may comprise both buried seed or seed in fire-responsive seed follicles or cones (see Chapter 16).
(c) Colonisation by wind-dispersed species. There are 'fire weeds' which will colonise a burned site for a few generations. These often give rise to a 'pyric succession' in which these species are eventually replaced by a slower growing vegetation type, often shrub and woodland communities. The succession represents a response to a combination of a nutrient flush and increased light. Typical species include the willow herbs *Epilobium* spp. (Onagraceae), and various composites and short-lived trees.
(d) Development of the pyric succession towards climax vegetation. In the most seriously phosophorus- and nitrogen-limited sclerophyll vegetation types, namely heath, bog, tundra and forest tundra, a pulse of nutrient availability always results from the fire. It is hard to be definitive on the precise cause-and-effect relationships because of lack of quantitative data and great variability of circumstance. One possible hypothesis is that the phosphate pulse is the more fundamentally important. This is probably true in Australian heath (see Chapter 16) and some tundra types where low decomposition rate causes a bottleneck in the phosphorus cycle. An increase in available phosphorus permits assimilation of any fire-mineralised nitrogen and may even accelerate nitrogen fixation or microbial mineralisation processes.
(e) Generation of a stable rhythmic system. In grasslands it must be remembered that the elemental contents of burnable senescent tissue are at a minimum. Burning thus leads to minimum loss and maximum probable mineralisation. A short fire cycle could be seen as part of a regular and stable cyclic system.

Part Two

Environmental complexes

The study of soil is dominated, for historic and practical reasons, by two main trends. On the one hand, agriculture has demanded knowledge directly applicable to crop production. On the other, earth-science-oriented studies, treating soil as the 'skin of the Earth', have led to the discipline broadly termed pedology.

A major problem in considering soil in an ecological context is where to start, in the sense of the ideas supported by the literature, and then when to stop. The position adopted here is that the most relevant unit of study is the rooting zone of natural vegetation. This gives licence to us to draw selectively on the strictly pedological or agricultural literature but to omit discussion of topics such as soil classification or soils as land resources for agriculture. Texts such as Russell (1973) or Foth (1978) present this material very effectively and comprehensively.

A soil may be regarded as a series of surfaces with complex properties permeated by aqueous solutions and gas mixtures which must be assumed to be in a constant state of variation. Soil condition varies horizontally, vertically and in time. In considering soil conditions relevant to plants and ecosystems it is of paramount importance to realise that even if we could describe or predict every single individual interaction, the whole system would be far from understood.

Soil is not only an extremely complex and relatively impenetrable portion of the general environment, but it is also the locale for important ecosystem processes. These include absorption and assimilation of mineral elements, herbivory and other forms of predation, and decomposition and consequent mineralisation. Soil is also the outer skin of the lithosphere, and earth chemistry is the main determinant of soil, and ultimately biological, chemistry. It is worth while to review the relative ratios of elements in the lithosphere and in organisms as a prelude to understanding what causes what.

7

Soil formation

All working ecologists with interests in vegetation find themselves at some stage in their careers drawn into close contact with the earth sciences of pedology, geomorphology and geochemisty. In the present context of understanding how soils form, the state factor equation of Jenny (1941), although later modified (Jenny 1961), offers a reasonable and straightforward framework for organising ideas. It is emphasised that the equation is purely a symbolic expression of the interactions entailed in the formation of soils, and is without a solution in the mathematical sense. The equation reads:

$$s = f\,(cl, or, r, p, t \ldots)$$

where s is any soil property, cl is climate, or is organisms, r is relief, p is parent material and t is the duration of the soil-forming process. Refinement of this verbal model is probably not very helpful to ecologists who intuitively accept the idea that ecosystems and environment are in dynamic equilibrium. We can also assume that soils are open systems with inputs and outputs of materials and energy. Although it is emphasised that interaction is the dominant concept and separation of factors may not be possible in practical terms, it is worth briefly reviewing each of these factors separately.

PARENT MATERIAL

A broad view of the lithosphere (Table 7.1 and Figure 7.1) using data from Bowen (1979) shows that 99% of the Earth's crust is composed of oxygen and silicon (75%), followed by aluminium, calcium, iron, sodium, magnesium, potassium, titanium and

Figure 7.1: Key differences between the elemental composition of (a) the Earth's crust and soils and (b) vegetation. Pie charts are based on the data in parts per million

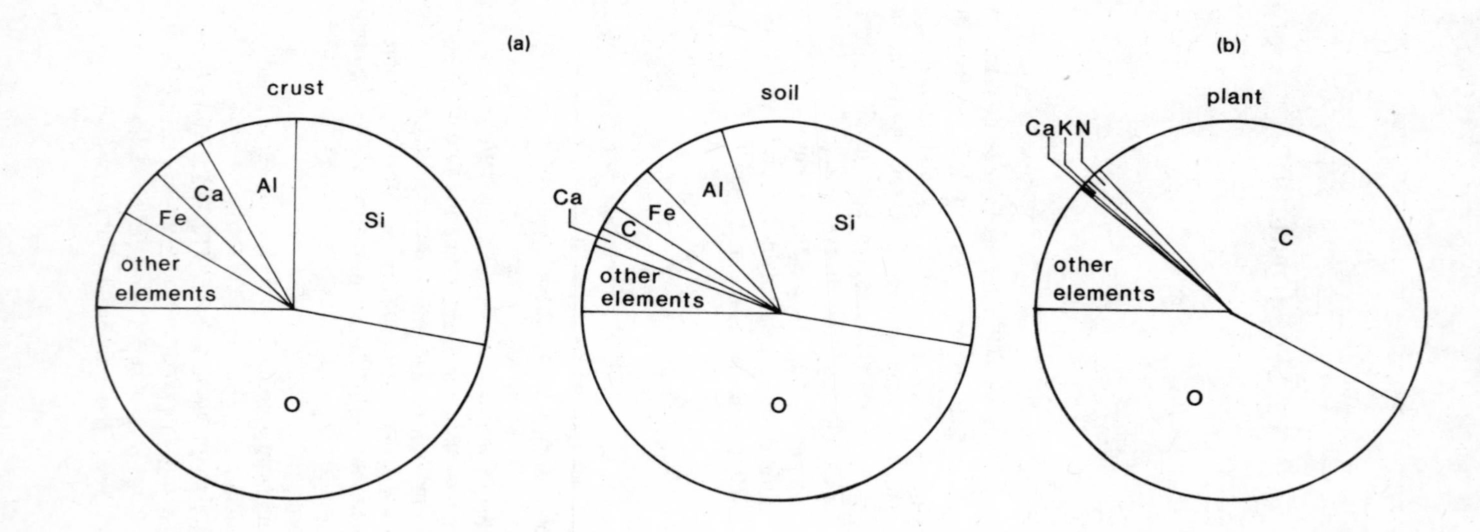

Table 7.1: A comparison of the elemental composition of the Earth's crust, soils and plant tissue. Data from Bowen (1979), expressed in milligrams of the element per kilogram

Element	Crust (median)	Soil (median)	Plant (range)
Oxygen	474 000	490 000	397 000–440 000
Silicon	227 000	330 000	200–62 000
Aluminium	82 000	71 000	38–530
Calcium	41 000	15 000	400–13 000
Iron	41 000	40 000	2–700
Sodium	23 000	5 000	2–1 500
Magnesium	23 000	5 000	700–9 000
Potassium	21 000	14 000	1 000–68 000
Titanium	5 600	5 000	<0.02–56
Phosphorus	1 000	800	120–10 000
Fluorine	950	200	0.02–24
Manganese	950	1 000	0.3–1 000
Barium	500	500	4.5–150
Carbon	480 (inorganic)	20 000 (organic)	450 000
Strontium	370	250	3–400
Sulphur	260	700	600–8 700
Zirconium	190	400	0.48–2.3
Vanadium	160	90	0.001–2
Rubidium	90	150	0.5–52
Nickel	80	50	0.02–5.3
Zinc	75	90	1.4–400
Cerium	68	50	0.01–16
Copper	50	30	4–20
Neodymium	38	35	0.3–7
Lanthanum	32	40	0.003–15
Nitrogen	25	2 000	12 000–75 000
Cobalt	20	8	0.005–4.6
Lead	14	35	0.2–20
Boron	10	20	8–200

phosphorus. The processes of soil formation lead generally to the removal of some sodium, magnesium and calcium which are dominant cations in sea water (see Figure 18.3). Carbon and nitrogen are added from biomass. In contrast, silicon, aluminium and titanium are virtually excluded from plants (Figure 7.1b), but 50% of plant tissue comprises carbon and nitrogen, ultimately from the atmosphere.

Coming from a broad view of the lithosphere to a consideration of individual rocks, it is clear that the physical fabric of soil as well as chemical characteristics depend largely on rock type. Even so, the common geological classification of rocks as igneous, sedimentary or metamorphic is not especially helpful to the ecologist. The important characteristics are:

(a) elemental composition of rock;
(b) rate of release of ions into the environment by weathering;
(c) particle size range produced by weathering;
(d) depth of weathering with respect to potential biological activity.

There are many examples where changes in rock types, even in the absence of alteration in relief or climate, are strongly and sharply correlated with ecosystem type. Perhaps most striking and widely known are the contrasts between ecosystem types on limestone and those in adjacent non-calcareous rock types (Chapter 19).

Of contemporary interest is the widespread observation that acid soils associated with siliceous rocks are susceptible to acidification by acid-polluted rainfall, e.g. Braekke (1976) and Harper (1982). At a more local level, in Australia the presence of nutrient-poor deep sands with a complex recent geological history has given rise to a series of heathland communities in the place of the characteristic woodland type for the region, e.g. 'wallum'-type heath instead of rainforest in Queensland, or heath replacing dry sclerophyll forest in Victoria (see Chapter 16). Still more localised, but relatively frequent, are geochemical anomalies, often associated with metamorphism. Grassy 'balds' on mountains (Kruckeberg 1984) and other communities of low productivity are associated with serpentine (Procter and Woodell 1975) (see Chapter 19). The survival of relict arctic–alpine plant communities at relatively low altitude in upper Teesdale is attributed to mineralisation by lead of limestone metamorphosed by an igneous intrusion (Jeffrey 1971; Chapter 13).

There is a large literature on base metal anomalies and closely correlated floras (Antonovics, Bradshaw and Turner 1971) giving rise to the development of biogeochemical methods of prospecting (Brooks 1979; Chapter 14).

CLIMATE

Whereas the influence of geological substrate on soil and vegetation is frequently clear, it is less easy to separate the effects of climate on a soil factor from the direct effect of climate on vegetation dynamics or other aspects of ecosystem function. For example, the process of accumulation of organic soils giving rise to bogs and fens (paludification) in North America and western Europe is heavily dependent on climate (Moore and Bellamy 1973). In this example 'climate' means the balance between precipitation and evaporation combined with a cool to cold temperature regime. Under these conditions decomposition of plant material is inhibited and organic material accumulates. The vegetation pattern in wetlands is frequently complex, with secondary effects of landscape topography, water chemistry, and hydrologic flow superimposed on the climatic regime.

In spite of this kind of complication, it can be taken for granted that soil-forming chemical reactions are accelerated by rises in temperature. A doubling of reaction rate for each 10°C rise in temperature is a crude approximation. Removal of reaction products by leaching maintains reaction rate. Hence reactions such as the dissolution of calcite in limestone, the breakdown of feldspars in granite or the release of aluminium from aluminosilicates and its subsequent leaching, will all proceed at optimum rates under sustained high temperatures and high rainfall. There is thus a geography of soil types dependent on this kind of phenomenon; the term 'weathering' is very literally applicable in this context. The distribution of major units of soil classification, the great soil groups, are largely determined by climatic factors that on the whole transcend the effects of the parent material.

RELIEF OR LANDSCAPE

It should not surprise any ecologist to discover different soils in different landscape units. This may be due to a large number of causes, all of which emphasise the close relationship between geomorphology and soil characteristics. These relationships have recently been selectively reviewed by Gerrard (1981), who, while virtually ignoring effects of, or on, living systems, discusses examples familiar to ecologists.

The features listed in Table 7.2 may be divided into two sets.

(a) Those in which subtle differences in soil materials are correlated with topography. For example, these include the variations in a post-glacial landscape in which bedrock or a basal till is overlain by a pattern of assorted materials such as esker gravels, loessic silt loams, or the moulded-clay form of drumlins.

(b) The more complex situation where variation in topography generates a difference in soil-forming process. An obvious example is the situation in the bottom of a valley where a combination of accumulated fine material from valley sides, plus abundant water, gives rise to ill-aerated and 'gleying' soils (see Chapter 9). It will also be seen that processes related to biological effects on soil formation are connected with aspect, altitude and slope. The tundra landform is discussed in Chapter 17, and a scheme illustrating soil differences in saltmarsh zones is set out in Chapter 18.

Table 7.2: Landscape and topographic features contributing to soil development

(1) *Correlations between parent material and topography:*
- (a) particle size range differentiation due to sorting, transport and deposition processes, e.g. glacial and fluvioglacial landscapes, coastal landforms, river floodplain, loess distribution;
- (b) correlations between solid geology and topography, e.g. the capping of a limestone upland by sandstone or shale

(2) *Soil processes dependent on topography:*
- (a) soil mobility or stability, especially in uplands with steep slopes
- (b) soil water status, which relates to aspect, soil depth and texture
- (c) slope failure linked to slope angle and drainage may produce recurrent landslips
- (d) systematic changes in soil chemistry and morphology, mainly determined by slope, referred to as 'catenas'
- (e) topographically determined differences in drainage, especially giving rise to wetlands in low ground with impeded drainage

Soils may thus be mapped in the local landscape as series. Where available, local soil maps and their commentaries will certainly contain ecologically important information whether the map is prepared for agricultural, resource management, geological or planning purposes.

TIME

The duration of the soil-forming process is a factor that emerges occasionally into the ecologist's consciousness. This phenomenon can only be studied effectively when similar parent material is exposed for variable times to similar regimes of soil formation. The examples given in Table 7.3 demonstrate that a study of time sequences entails integrating all the other soil-forming factors. Biological activity is both a powerful cause and consequence on a long time series, with organic matter and nitrogen addition to a mineral substrate being a recurrent theme.

Table 7.3: Studies on the time scale of soil formation (see Stevens and Walker 1970)

Parent material	Authors
Seacoast dune sands	Many, e.g. Salisbury (1925), Burges (1960), Ovington (1960)
Inland sand dunes	Olson (1958)
Glacial moraine series	Many, e.g. Chandler (1942), Crocker (1952), Stevens (1963)
Volcanic ash series	Tezuka (1961), Ruxton (1968)
Alluvial soils	Jenny (1962)
Polder soils, formally sea covered	Ente (1967)

Another important matter to be revealed is the rate of processes of removal that are only hinted at by short-term studies. The two best examples are the decalcification of dune sands and the removal of phosphate from soil profiles. Decalcification is a relatively rapid process, with several studies indicating that sands with a total $CaCO_3$ content of less than 5% will lose it from the top 10 cm of surface within 400–500 years in the temperate zone (Ranwell 1972). This removal is in spite of recycling of blown material.

There have been fewer studies of phosphate removal, but Stevens and Walker (1970) conclude that the half-life of phosphate in temperate soils is in the order of 20 000 years. More chronosequence studies leading to this kind of perspective need to be carried out because of their relevance to long-term changes in vegetation cover on a continental scale. However, current work is indicating that precipitation may contribute some ions to landscapes (Chapter 15).

ORGANISMS

While the main thrust of our thinking is to understand how soil influences vegetation and ecosystems, it should be apparent that there is a substantial reciprocal effect. Organisms cause additions to, and removal from, the soil fabric, and participate in processes of comminution and aggregation of particles. (See Table 7.4.)

Table 7.4: Soil processes consequent on biological activity

Addition: Carbon and nitrogen compounds, organic particulates
Removal: Water, nitrogen and carbon, oxygen
Comminution: Organic particles; inorganic particles from weathering leading to pH change
Aggregation: Polysaccharide production by bacteria and from mycelial threads and faecal pellets
Transformation: Many — see nutrient cycles; gleisation; silicate → silica (opaliths)

DEVELOPMENT OF A SOIL PROFILE AND SOIL CLASSIFICATION

The two great processes that combine to cause soils to form and develop are:

(1) rock weathering, implying all that has so far been described in chemical and physical effects;
(2) biological productivity, whereby carbon compounds are produced and the energy for nitrogen fixation is channelled to the appropriate organisms.

Both processes are governed by the climate of radiation flux and rainfall regime. In a given site this interaction generates a soil profile between organic production at the soil surface and weathering at the rock head. A soil profile may be described in terms of the distribution of organic matter initially undecomposed, a weathered layer and unweathered parent material. Figure 7.2 shows a simple example.

A widespread auxiliary process is the vertical transport within the profile of particulates or substances in solution. Most commonly this means the downward movement of materials following gravitational drainage of water, i.e. leaching. Leaching will tend to remove fine particulates, ions and organics in solution down the profile where they may reaccumulate (e.g. iron salts or calcium carbonate) or ultimately move into groundwater (e.g. nitrate ions). Against leaching are the extraction of ions by plants and the bioturbation activities of soil organisms, best exemplified by the earthworms.

Climatic situations leading to high evaporation rates at the soil surface and upward mass movement of groundwater are also widespread and important determinants of vegetation type and potential land use.

Bearing in mind these interactions, which may occur to generate an individual soil profile, it is to be expected that regional patterns occur on a world scale of major soil groups. The problems of identification and classification are made more formidable because of the different perspectives and needs of different groups of scientists:

(a) Pedologists: geomorphologists interested in the weathered surface of the Earth's crust as a phenomenon to be studied in its own right. This includes both the profile development aspects and the geographic aspects. It may be argued that because they have the widest geographic perspective, namely the whole Earth's surface and even the surface of other planets, their views on soil classification are of greatest value.
(b) Agriculturalists or land resource utilisation specialists: their aim is the practical use of the Earth's land surface for crop production, whether that be of arable or forage crops, for forestry or for the utilisation of essentially natural ecosystems. They need means for analysing and interpreting profile descriptions and for understanding the dynamics between soils and cropping systems.
(c) Ecologists: the ultimate preoccupation of ecologists is studying problems at the ecosystem level. Their need for soil classification ranges from a rather casual inclusion of a diagnosis of soil type in a site description to the study of soil formation as an integral part of an ecosystem analysis, e.g. of a dune, saltmarsh or bog.

There are no universally useful or internationally acceptable schemes for soil classification, and many national systems exist. Where a suitable local system does not apply, a line of least resistance on soil classification to be adopted by ecologists may be to accept for the time being the latest approximation of the United States Department of Agriculture. This system, which depends at the finest level of resolution on knowing the developmental history of a soil type, has a sound intellectual basis. Its taxonomic system resembles that of the Linnean system for species or Braun-Blanquet's system for plant communities. The most unfortunate aspect of it is the terminology applied to lower orders. These make no room for other familiar terminology and consist entirely of neologisms unpronounceable in any language. At the finest level of resolution, which frequently means describing the comparative differences or similarities in soils associated with subtle ecological

Figure 7.2: Schematic view of a simple soil profile showing horizon development

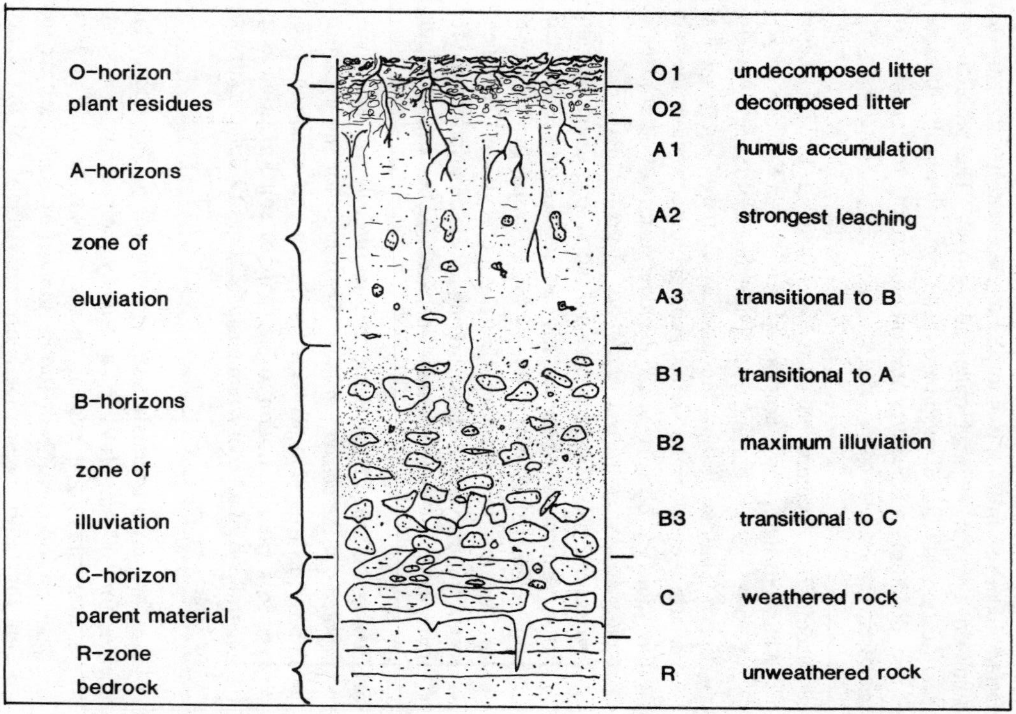

O-horizon plant residues

- O1 undecomposed litter
- O2 decomposed litter

A-horizons zone of eluviation

- A1 humus accumulation
- A2 strongest leaching
- A3 transitional to B

B-horizons zone of illuviation

- B1 transitional to A
- B2 maximum illuviation
- B3 transitional to C

C-horizon parent material

- C weathered rock

R-zone bedrock

- R unweathered rock

Figure 7.3: Diagram to integrate the processes of soil formation emphasising their interactive nature

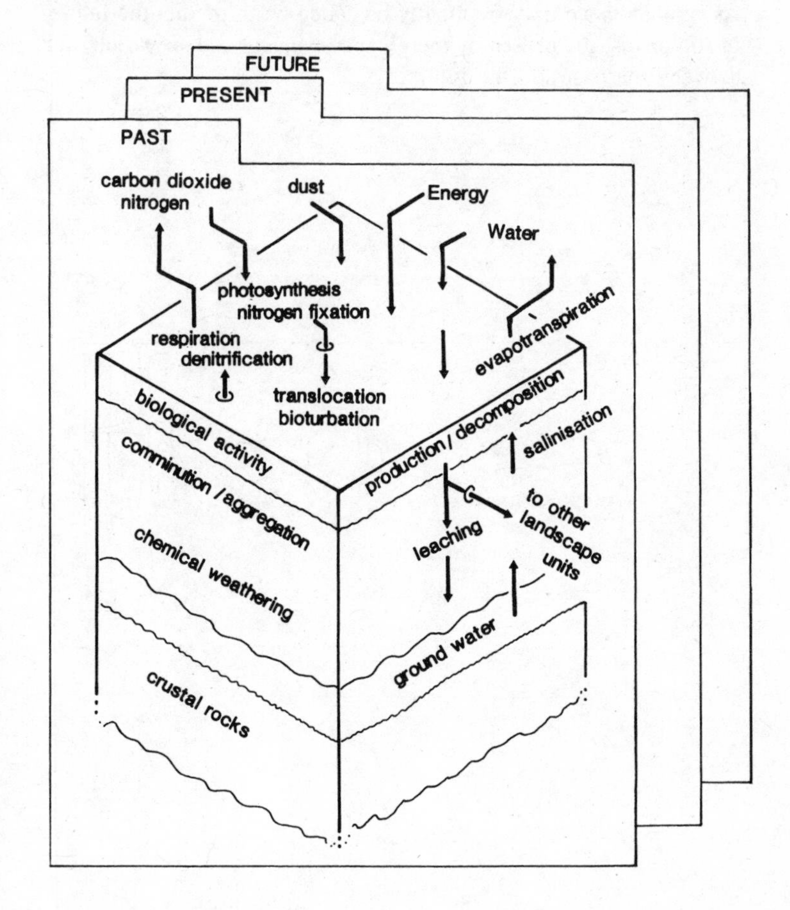

difference, the USDA system, or any other, may have little or no relevance. The portion of the profile exploited by roots or other biota must then be described in ecologically meaningful terms that need have little or no bearing on soil formation.

Figure 7.3 has been prepared to symbolise the major processes of soil formation as an aid to integrating the concepts involved. It is important to be flexible in one's thinking about soil development, especially where size of unit is concerned. Soil changes can occur within centimetres horizontally and vertically. Determining if these are of an ecologically significant amplitude is a recurrent theme. Similarly rapid cyclic and progressive changes may occur with time. As ecologists we may frequently be called on to predict the future. In soil terms, the present is merely a convenient and accessible link between the past and the future.

8

Soil matrix and soil water

PARTICLE SIZE CLASSES

The most obvious and naive model of soil particles is of glass marbles piled in a jar, rigid, rounded and regularly sized particles which have a regular series of pores permeating the mass. The geometry of sphere packing (Figure 8.1) shows that as the spherical particles get smaller, the surface area of the system increases and the size of the interstitial spaces decreases. The total interstitial volume of such systems trend towards 20–25% of the total.

The only real system which in any way resembles the spherically shaped and uniform-size glass marble model is that of coastal sands. The generality of soils are composed of particles which range from gravel (diameter > 2 μm) to clay (diameter < 2 μm). Soil particles are not usually spherical, smooth or of a single size class. Physical and chemical weathering will reduce rocks to angular particles of a wide size range. Only further weathering in aquatic environments, and to some extent by wind transportation, will round and sort particles. We accept that in explaining and describing the structure of soil we are dealing with a continuum of particle sizes that may, if only for practical purposes, be broken down into convenient size classes. There is only rather broad agreement about how to do this internationally. The most constant point of agreement is that particles larger than 2 mm in diameter are to be disregarded for analytical purposes. As a 2 mm mesh sieve is used to separate them in a sample collected from the field, such particles may be called 'coarse tail'. All particles passing a 2 mm sieve are collectively termed 'fine earth'. The larger particle sizes (> 2 mm diameter) are frequently ignored in the agricultural literature, as stony soils hinder arable cultivation and the presence of stones is not thought to be positively relevant to

Figure 8.1: Size, packing and porosity of soil particles

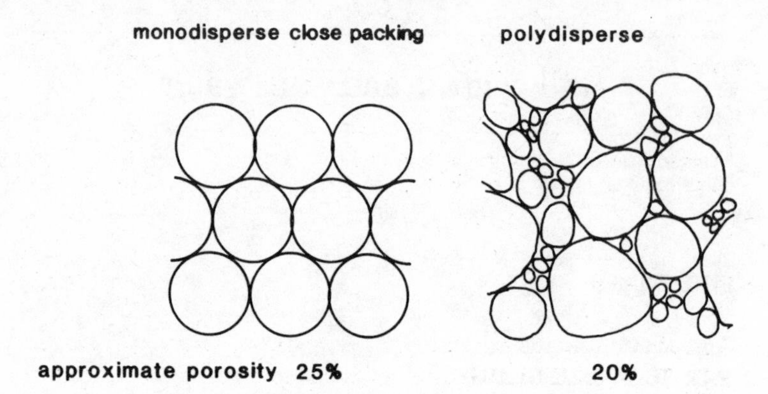

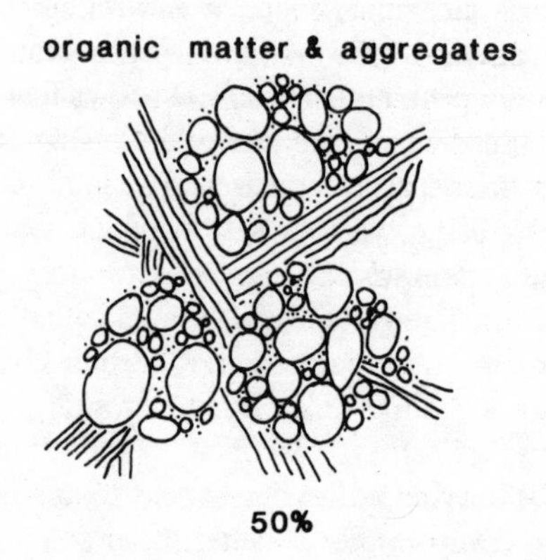

crop growth. The ecologist should not ignore them, however, as stones are substrates for bryophytes, lichens and algae, and have a role in controlling evaporation, soil permeability and the interception of water by precipitation and dew formation. It is convenient when describing a soil or sediment as habitat to express the weight of 'coarse tail' as a percentage of the total 'fine earth'. This value may frequently exceed 100%.

Table 8.1: Classification systems for the fine earth fraction

	Size (μm)		
	Britain	USDA[a]	ISSS[b]
Sand	2000–60	2000–50	2000–20
Silt	60–2	50–2	20–2
Clay	< 2	< 2	> 2

[a] United States Department of Agriculture.
[b] International Soil Science Society.

FINE EARTH

Agreed terms for the fine earth fraction are sand, silt and clay (Table 8.1). There is some practicality in adopting 50 or 60 μm diameter as the upper size limit for silts. First, it is possible to use a sieve to separate these sizes, and secondly, this particle size classification correlates strongly in temperate regions with agriculturally useful textural classes (Russell 1973). If we take for granted the need for a descriptive classification of particle size and accept a simplification into sand–silt–clay classes, then data may be compared by entering values for each sample as a single point on a triangular diagram (Figure 8.2).

If more particle size data are available for interpretation, then a more satisfactory system is to plot a cumulative curve of the percentage contribution of each particle size class against a convenient logarithmic scale of particle diameter. An example is given in Figure 8.3 of a series of soils from a saltmarsh. The three curves merely point towards differences in soil porosity which may or may not contribute to observed ecological pattern.

A particle size determination will go some way towards indicating testable hypotheses about soil structure, cation exchange capacity and water-holding properties, as well as about the parent material and development of a soil. The properties of particle size fractions cannot be understood without at least a brief excursion into their mineralogy. This is very greatly simplified here, and reference could usefully be made to geological texts.

The reason why a mechanical analysis of particle size fractions is of only indirect help in interpreting soil as environment is that there are at least two levels of soil organisation. These are the results of the processes of (a) aggregate formation, and (b) bioturbation, or mixing by the burrowing activities of soil fauna. In both, the presence and turnover of organic matter is of paramount importance.

Figure 8.2: Triangular diagram for representing the fine earth of a given soil as a single point

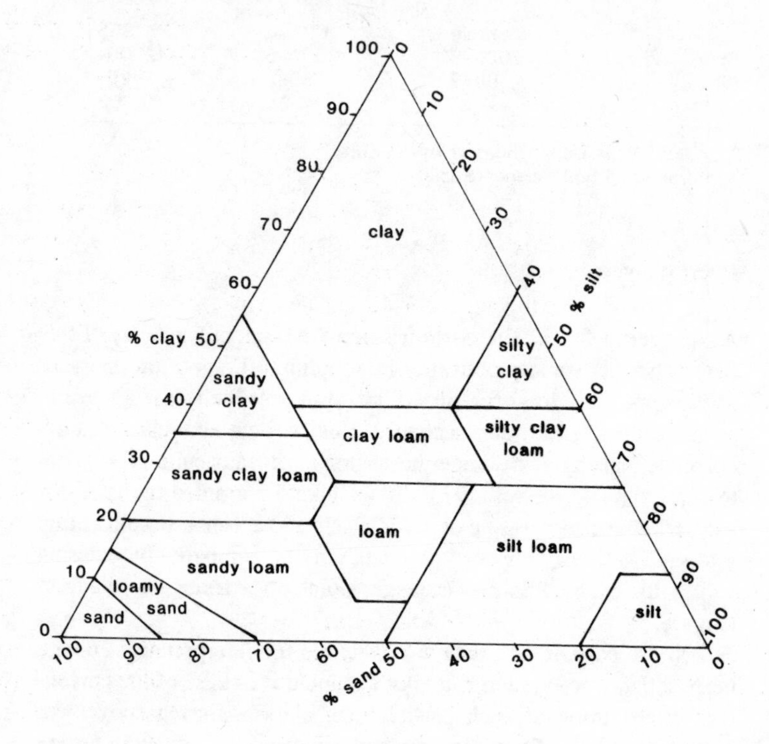

MINERALOGY OF SOIL PARTICLES

Sands and silts

From the inventory of elements in the Earth's crust (Table 7.1) it is seen that oxygen and silicon comprise 75%. The size of the atoms or ions that comprise the bulk of crustal elements indicates first that the 1.39×10^{-10} m radius for oxygen is 3.5 times larger than the 0.39×10^{-10} m for silicon (see Table 8.2).

The silicon atom can be easily accommodated between a tetrahedron of four oxygen atoms. Such a tetrahedron may include cations such as iron, magnesium or calcium. Because of the possible geometric arrangements of these tetrahedral clusters, oxygen sharing occurs, and a mineral series exists between a 4:1 O:Si ratio in

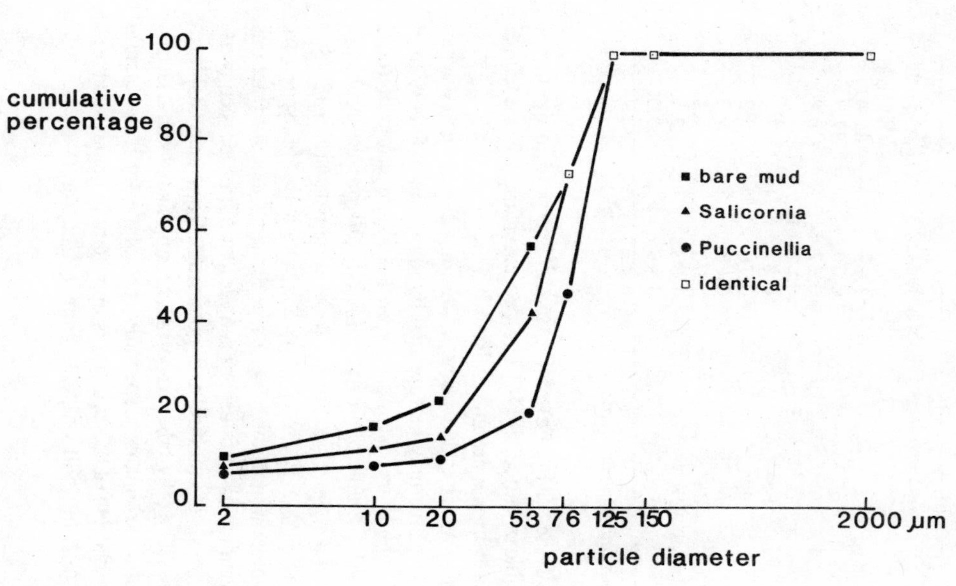

Figure 8.3: Cumulative curves of particle size make-up of three zones of a Lancashire (UK) saltmarsh

Table 8.2: Ionic radii of the most common elements in the Earth's crust as a guide for interpreting soil mineral structure

	Valency	Ionic radius, Å[a]	% of Earth's crust
Oxygen	O^-	1.39	47.4
Silicon	Si^{4+}	0.39	27.7
Aluminium	Al^{3+}	0.57	8.2
Ferrous iron	Fe^{2+}	0.83	4.1
Ferric iron	Fe^{3+}	0.67	
Magnesium	Mg^{2+}	0.78	2.3
Sodium	Na^+	0.98	2.3
Calcium	Ca^{2+}	1.06	4.1
Potassium	K^+	1.33	2.1
Ammonium	NH_4^+	1.48	0.002

[a] Å = Angstrom unit = 10^{-10} m.

the mineral olivine, $(Mg, Fe)_2 SiO_4$, to the 2:1 ratio in the three-dimensional quartz lattice, SiO_2. The decreased ratio is associated with an increase in weathering resistance.

In addition to an extended range of binding cations and oxygen sharing, silicon atoms may be replaced by aluminium. This is so in the feldspars, which are the most important single group of rock-forming minerals. Either one or two in four silicons are replaced, e.g. potassium feldspars ($KAlSi_3O_8$), sodium feldspars ($NaAlSi_3O_8$), and calcium feldspars ($CaAl_2Si_2O_8$). Silica and the feldspars are generally the most important minerals in the sand and silt fractions of soil. The weathering of feldspars is an important process converting elements such as calcium and potassium from unavailable to available forms. The only other important group of sand–silt–fraction minerals are the carbonates, especially the calcite form of calcium carbonate. Weathering of this mineral is important in maintaining high pH in some soils and will be dealt with in Chapter 19.

Clays

The weathering of primary minerals gives rise to ions which may recombine as fine crystalline particles, typically smaller than 2 μm in diameter. There are two groups of clay minerals, the silicate clays and the oxide clays. The latter, which are primarily finely divided iron and aluminium oxides, are of major importance in the humid tropics. The silicate clays are widely distributed and are important in an understanding of the physical structure and cation exchange properties of soils.

Formation of silicate clays

Silicon–oxygen tetrahedra can form as sheets, with alternate units inverted. Such a sheet may be represented as $Si_2O_5^{-2}$. Two of the apical oxygens are unsatisfied and therefore can attract cations plus enough additional anions to neutralise all the cation charge. If the $Si_2O_5^{-2}$ unit attracts an Al^{6+} ion, then four OH ions will also be attracted. This dual sheet may continue to crystallise out of soil solution with a repeat of each layer at about 7.3 Å (Figure 8.4). Sheets are held together by hydrogen bonding of hydroxyl oxygen. This is the clay mineral, kaolinite. The formation of kaolinite may occur as a result of weathering from a feldspar; here it is formed as a result of recrystallisation. Clay minerals in general may originate from either weathering or recrystallisation, but usually under different circumstances.

The example of kaolinite is chosen not because it is a common clay, but merely because it is the most simple possible case with 1:1 ratio between silicon tetrahedra and aluminium octahedra. A more common case is the 2:1 ratio with two silica sheets enclosing an alumina sheet in sandwich form. Further variation occurs when substitution of atoms in the crystal occurs. These can include aluminium for silicon in the outer sheets of the sandwich in illite. This gives rise to additional negatively charged sites for cation exchange. Aluminium in the core of the crystal may also be replaced by iron or magnesium, again giving rise to sites for cation exchange.

The characteristics of the more common clay minerals are set out in Table 8.3. It should be noted that the tendency for ion exchange will depend on the type of clay as well as on the quantity in a soil.

SOIL AGGREGATE FORMATION

An aggregate is a cluster of particles held by some form of bond but brought together by physical forces. Aggregates produced vary greatly in size. At the top end of the scale are columnar blocks several centimetres in diameter produced by the contraction and fracturing of a contractile-clay-rich soil on drying. This may lead to cracks of the order of a centimetre in width appearing between such blocks. At the other end of the aggregate scale are the largely organic faecal pellets of microarthropods and the casts of annelids, which may be largely of mineral material. In both these cases the aggregate is stabilised by biological polymers. Furthermore, aggregate formation and stability are generally aided by microbial

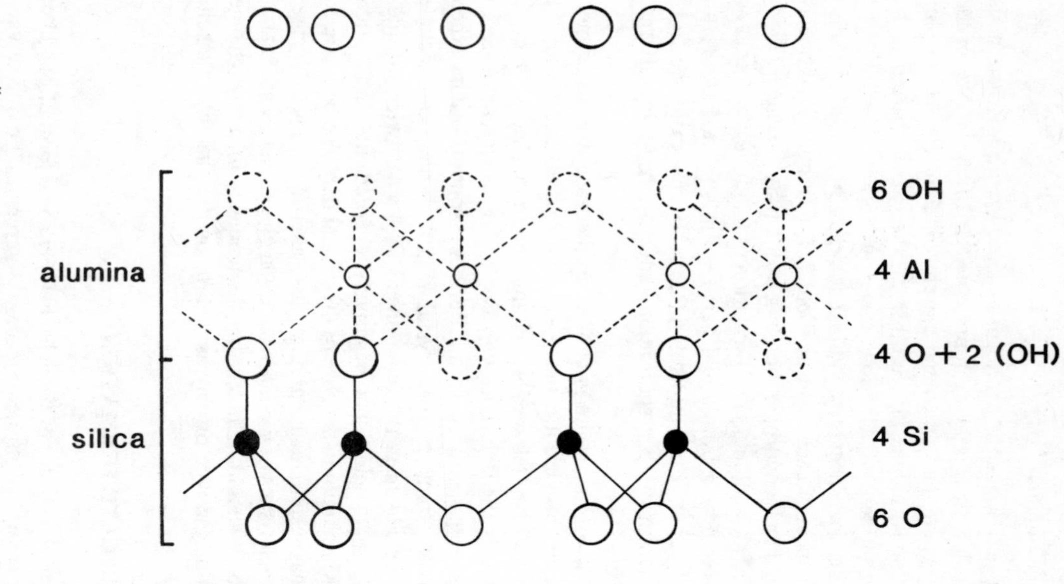

Figure 8.4: Structural diagrams of some clay minerals. The atomic structure of the layers of kaolinite illustrates the sharing of oxygen atoms by silica and alumina sheets which is characteristic of clay minerals. Some possible arrangements of sheets are shown for three particular minerals, whose properties are outlined in Table 8.3

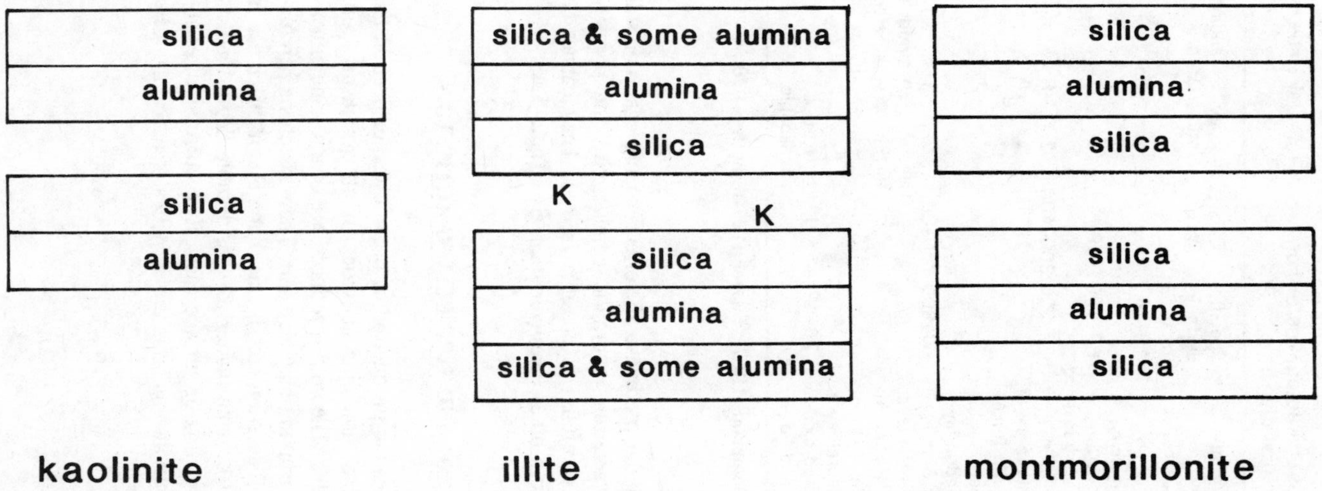

silica
alumina

silica
alumina

kaolinite

silica & some alumina
alumina
silica
K
K
silica
alumina
silica & some alumina

illite

silica
alumina
silica

silica
alumina
silica

montmorillonite

Table 8.3: Chemical characteristics of some clay minerals; see Figure 8.4 for structural diagrams

Clay mineral	Substitution or other characteristics	Specific surface ($m^2\ g^{-1}$)	CEC (meq $100\ g^{-1}$)	Lattice ratio	Expanding lattice
Kaolinite	Little or no substitution	10–20	8	1:1	No
Montmorillonite	Mg for Al in octahedral layer	700–800	1000	2:1	Yes
Illites (clay micas)	Al for Si in tetrahedral layer. Fe or Mg for Al in inner octahedral layer	100–300	200–400	2:1	No (K bridges)
Vermiculite (derived by weathering from clay micas)	Al for Si in tetrahedral layer. Fe or Mg for Al in inner octahedral layer	750–800	750–800	2:1	Yes (K bridges weathered out)
Oxide clays	Hydrated oxides of Al and Fe	25–42	5–1	Not applicable	

Source: Data from Thompson (1957); Russell (1973); Foth (1978); Talibudeen (1981).

hyphae and extracellular products. The effects of aggregates on soil properties such as aeration, hydraulic conductivity and porosity must be measured carefully, if possible *in situ*. They cannot be predicted readily from other characteristics. See Table 8.4.

BIOLOGICAL MIXING AND TUNNELLING ACTIVITIES

This process includes the production of organic matter deep in the soil by roots and the tunnelling activities of all organisms from annelids to vertebrates. The voids produced are of the same dimensions as those of a shrinking-clay-rich soil. Their effects on percolation of water and soil aeration are similar, but a major difference is the incorporation of organic matter to considerable depth, and its complement the formation of surface casts. The burrows of earthworms are also important voids for the activities of other members of the soil fauna.

Table 8.4: Soil aggregate formation and stabilisation

(a) *Forces leading to aggregate formation*
Expansion of water in freeze–thaw cycles
Shrinkage and dilation of clays in wetting–drying cycles
Formation of faecal pellets and casts by soil fauna
(b) *Components of aggregate stability*
Capillarity of water–hydrogen bonding between water molecules and soil particles
Divalent cations. Aggregates between clay particles are held by Ca^{2+} bridges between adjacent platelets. Disaggregation can be induced if flooding with monovalent cations occurs, e.g. sodium
Colloidal oxides and hydroxides of iron and manganese
Humic colloids
Exuded polysaccharides of plant or microbial origin
Microbial filaments
Mucus produced by soil fauna
Fibrous organic matter

SOIL ORGANIC MATTER

Background

Soil organic matter can comprise from virtually zero to almost 100% of the soil fabric and it must be recognised as an essential complement to the mineral particles. The organic matter recognised in a soil at a given moment is the resultant between processes of production, both primary and secondary, and decomposition. Thus the range of organic molecules present in soil is very large, including products of secretion and excretion of living organisms, newly shed and post-decomposition residues of dead parts of organisms and whole organisms. In addition to identifying metabolites in trace quantities, analytical chemistry of soil organic matter has explored the nature of residual colloidal polymers, the so-called humic fractions. This literature does not seem to be moving towards meaningful ecological interpretation and will not be considered here.

There are several functional aspects to the role of organic matter in the soil environment that are best dealt with under other headings:

(1) fibrous and colloidal substances contribute to soil fabric;
(2) it stabilises soil aggregates;
(3) cation exchange capacity is associated with cellulosic materials;
(4) it adds to matrix water storage capacity;
(5) it plays a role in ligand binding of metals;

Figure 8.5: Bulk density of soil seen as a function of organic content

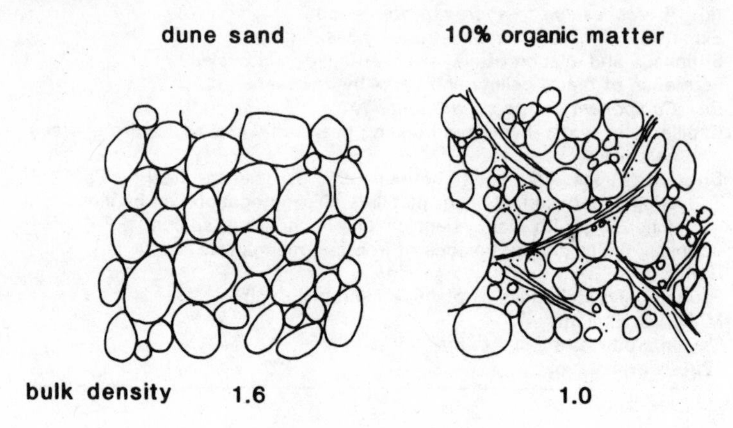

90% organic matter

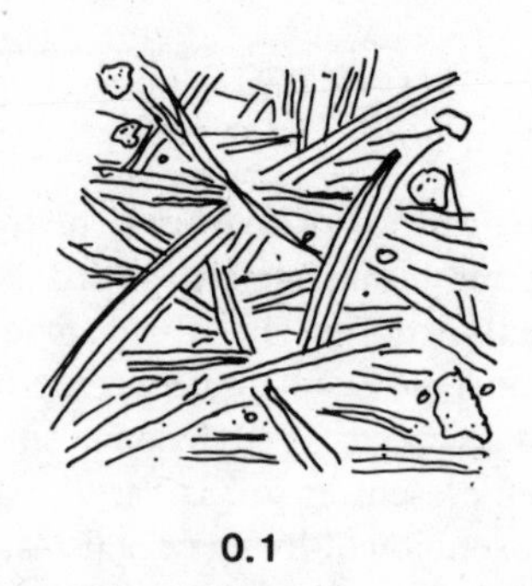

0.1

(6) it acts as a reservoir of nutrient elements, e.g. nitrogen, phosphorus, sulphur;
(7) it is a substrate for microorganisms and soil fauna;
(8) it plays a part in the reduction of the rate of heat flow.

Soil organic matter and bulk density

Soil is composed of components with very different specific gravities. That of quartz and other soil minerals is in the order of 2.6 g ml^{-1}, whereas soil organic matter may be in the order of 0.1 g ml^{-1}. The bulk densities of soils collected carefully from the field vary over the range 1.6 g ml^{-1} for dune sands to 0.1 g ml^{-1}

for scarcely humified sphagnum peat. Although the organic content of a soil is not the sole determinant of its density, it has been shown to be a reliable empirical predictor.

Realisation of the recorded range of bulk density and its strong dependence on organic matter content leads to two conclusions. First, in terms of soil function, organic content has a distinct bulking-up effect, which may also be thought of as a dilution of the elements present in the Earth's crust and soil.

Secondly, in terms of comparing soil analytical data, the futility of comparing dissimilar soils on a weight for weight basis is clearly seen. Common sense dictates that information be expressed on a weight for soil volume basis, e.g. milligrams of elements per litre (dm^{-3}) of soil. This has the dual rationale of correcting for density variation and acknowledging formally that plant roots, and other organisms for that matter, recognise and explore soil volume. The arithmetic of this conversion is to multiply the weight for weight data by the bulk density. Since bulk density is both difficult to measure accurately and is seldom included in published data, a useful aid has been devised in the form of a correlation between soil bulk density and loss on ignition of the soil, a crude index of soil organic matter which is convenient to determine (Jeffrey 1970; Harrison and Bocock 1981). Figure 8.6 illustrates this relationship as broadly applicable to all soils so far examined. For more precise correlations, a calibration curve should be prepared for particular sets of soil.

Production and transformation of organic matter

The production of organic matter and its transformation are described in Figure 8.7. If some organic matter is the result of current photosynthesis, some is being produced as faecal pellets of microarthropods and some is in the form of chemically protected residues, then a series of differing ages may apply to various fractions. The turnover of organic matter may be explored by means of various techniques. Short-term decomposition may be explored by means of the simple litter bag or description of $^{14}CO_2$ release from ^{14}C-labelled litter. The age of residues may be determined by the standard radiocarbon method of chronology. The conclusion of the studies presently available for agricultural soils is that rapid decay takes place in the order of months; various degrees of protection increase turnover time to some tens of decades; but a residue, amounting possibly to 50% of soil organic matter, will turn over in 1000–2000 years (Jenkinson 1981). The

Figure 8.6: A curve illustrating the relationship between soil bulk density and organic matter content (loss on ignition)

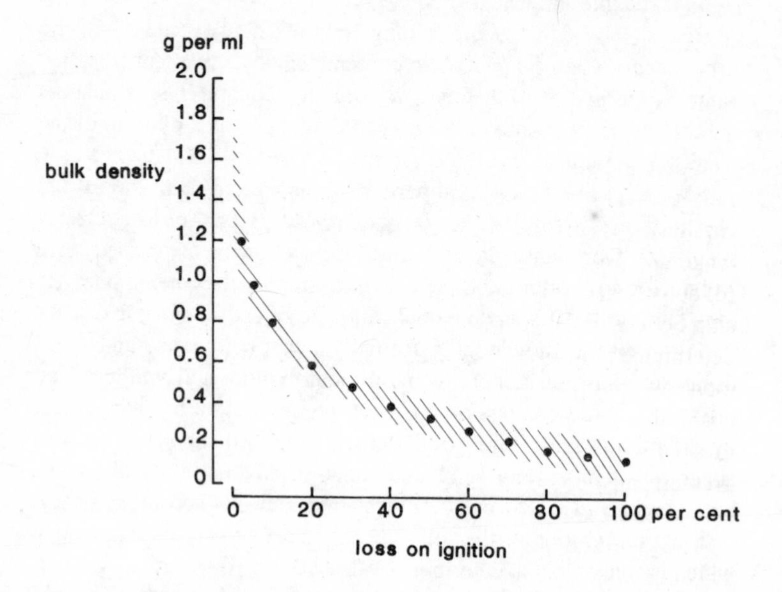

Figure 8.7: Production of organic matter and its transformation

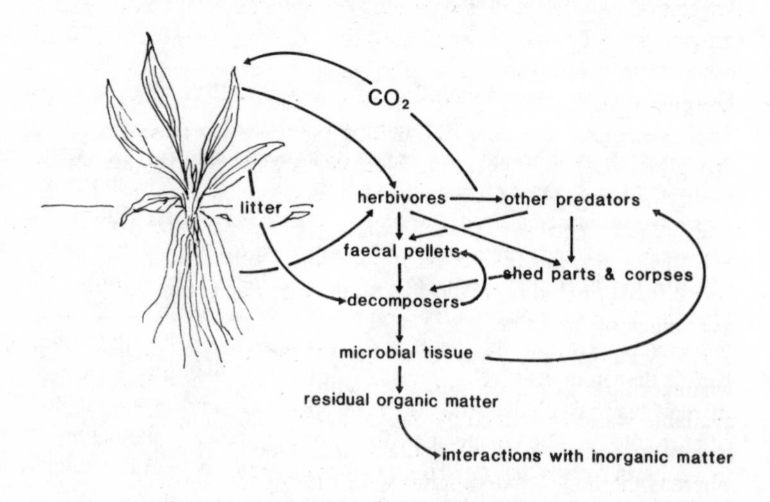

stability of this latter fraction emphasises its contribution to soil fabric. More work needs to be done on this aspect of soil organic matter in various ecological contexts.

SOIL WATER

The predominant force restraining water in soil is the matrix potential, ψ_m. The soil matrix may be thought of as a hierarchy of capillaries; its precise characteristics will depend on (a) particle size make-up; (b) organic matter content; (c) nature of aggregates; and (d) activities of soil fauna. Because of this complexity, the relationship between plant available water and soil water content should be determined for each soil individually. Such a curve, the soil moisture characteristic curve, is essentially a calibration curve enabling a gravimetric determination of soil water to be converted to ψ_m (Figure 8.8). Such a calibration may be carried out using a pressure membrane or plate apparatus (Figure 8.9) or by thermocouple psychrometry (Figure 3.2).

The key characteristic of soil water is the quantity of plant available water in the context of the water use by the vegetation. Plant available water is defined in Figure 8.8 as the quantity in a soil existing between the field capacity and the wilting point of the vegetation concerned. Field capacity defines the water held against gravity, any additional water being lost by drainage. The usual ψ_m assigned to it is -0.03 MPa. The value for ψ_m at wilting point is less easy to define, and a 'wilting zone' is sketched into the diagram. This symbolises the range of wilting-point values encountered in droughted natural vegetation. Most xerophytic species are sufficiently sclerophyllous to make the concept of wilting barely relevant. For agricultural use, wilting point is assigned a value of -1.5 MPa, and this is selected as one limit. The lower limit selected is -2.0 MPa. It is probable that some vascular plants will wilt at ψ_m values of less than -1.0 MPa, whereas others will survive at -2.5 MPa to rehydrate. However, the diagram indicates that whatever ψ_m value delimits the wilting point, the quantity of available water is defined by the soil. In the diagram the clay loam is seen to contain 31 g of available water per 100 g d.w. of soil, whereas the dune sand contains almost 16 g water per 100 g d.w. of soil. The value is estimated at field capacity in both cases.

Direct measurements of ψ_m may be determined by calibrated sensors emplaced in the soil. These range from gypsum blocks or

Figure 8.8: Soil moisture characteristic curves for a dune sand and a clay loam

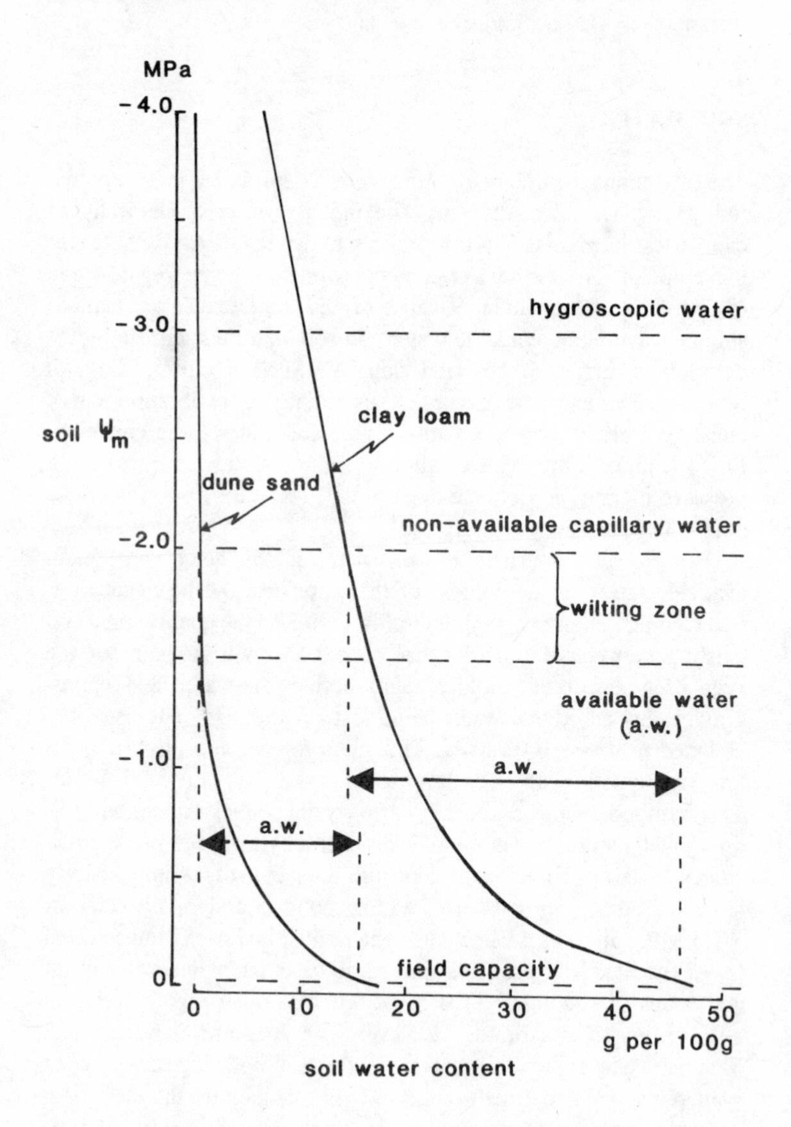

Figure 8.9: Pressure mer brane apparatus

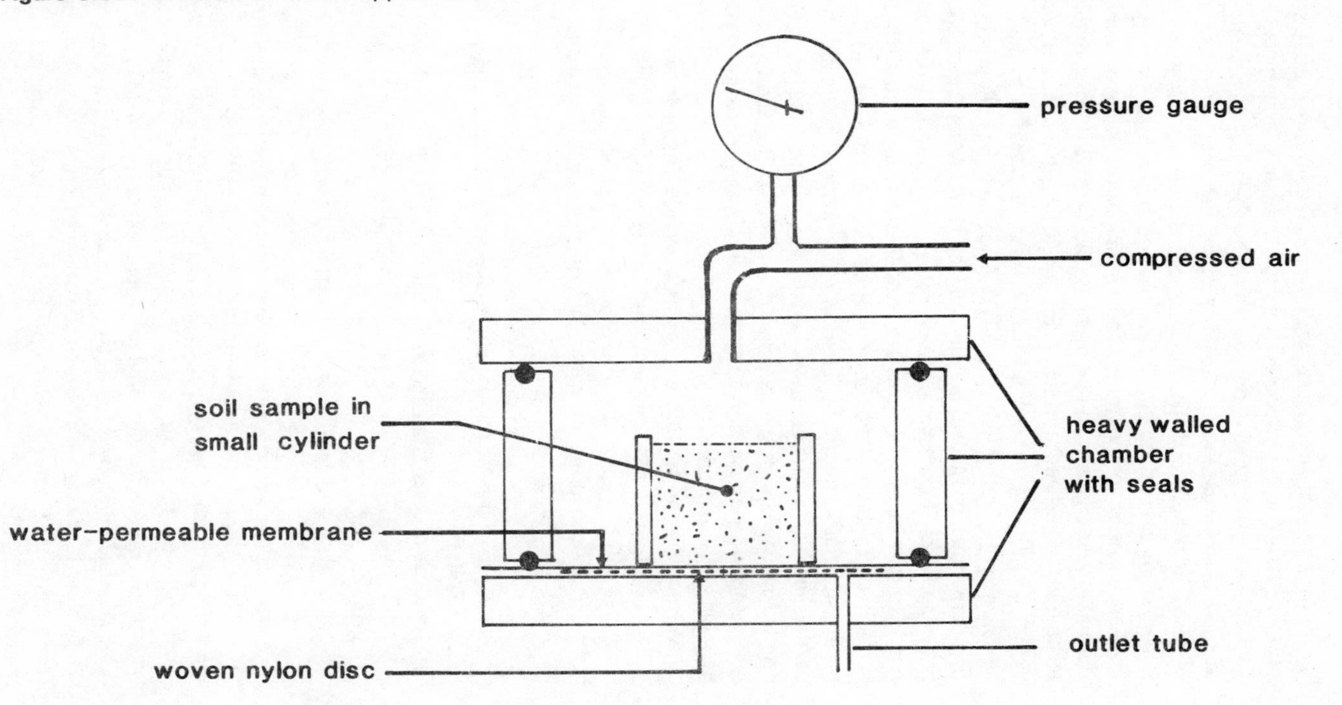

nylon pads in which conductivity is measured; osmotic cells with pressure transducers; psychrometers enclosed in porous pots; and vacuum gauges coupled to porous pots. The shared disadvantages of all these systems are the need to displace soil for installation, the cost of instrumentation, and the problems of reliable performance over a period of months in the field. On the other hand, emplaced instruments offer an opportunity for continuous recording of soil ψ_m if suitable data-logging and interpretation facilities exist. Most of these techniques have been developed for the management of irrigation systems, and their use is limited in ecological research to date.

The origin of soil water is usually considered to be bulk precipitation, groundwater or occasionally surface flow. A source of water that may be of importance in many situations is 'occult' precipitation, i.e. dew or mist. Condensation of liquid water occurs when a surface cools below the dewpoint of the atmosphere with its particular relative humidity. Liquid water may condense within soil cavities under these conditions. Mists are suspensions of water droplets in air that is just about cooled to its dewpoint, and commonly develop in air masses forced upwards by rising ground and cooled by adiabatic expansion.

As the solvent for all ions, water clearly interacts with the nutrient cycle of individual plants and whole ecosystems. Details of examples will be explored in Chapter 15, but it should be mentioned here that the coupling of water and nutrient cycles includes the following phenomena:

(a) addition of ions to soils and ecosystems by bulk and occult precipitation;
(b) removal of ions from plants by leaf wash and stem flow;
(c) removal of ions down a soil profile, ultimately to the groundwater;
(d) removal of soil particulates by surface flow.

Water vapour in the air, as measured by relative humidity, may be assigned a ψ value equivalent (Figure 8.10). A convenient approximation for this equivalent is 1.4 (100 − RH) MPa. It will readily be seen that if air has an RH of 97%, this is equivalent to an osmotic or matrix potential of approximately −4 MPa. This value is outside the range for all tissue osmotic potentials, except for halophytes. It is also well beyond the range of available soil water. This means that a water potential gradient from soils and plant tissue exists at 97% RH.

Figure 8.10: Water potential of water vapour in air in relation to relative humidity

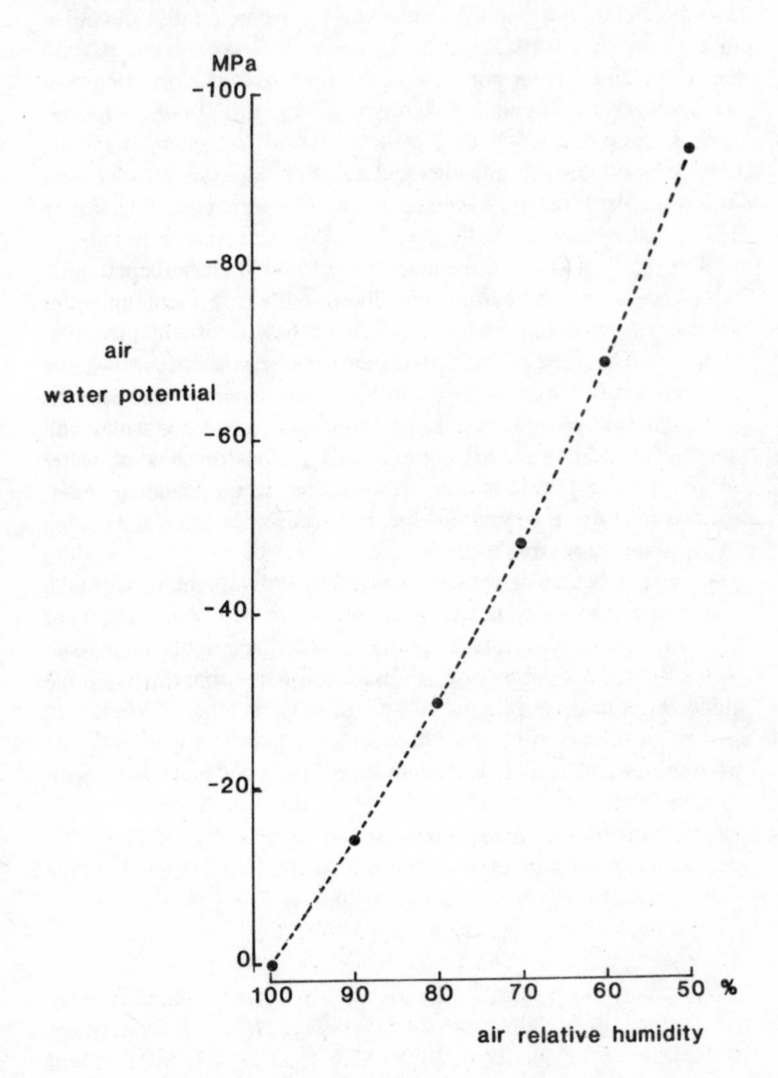

SOIL SALINITY

The ionic composition of the soil liquid phase is usually equivalent to ψ_s of −0.01 MPa or approximately 30% of the ψ_m at field capacity. Thus solute potential is usually a trivial term compared with soil matrix potential. The presence of high salt concentrations derived from sea water (ψ_s = − 2.4 MPa), or terrestrial saline deposits, reverses this situation and has profound ecological effects. These are explored with respect to coastal saltmarshes in Chapter 18.

The most satisfactory method of soil salinity measurement is by the thermocouple psychrometer, either with sliced 1 cm core sub-samples or expressed interstitial water. In practice it is not possible to use psychrometry in the field, and field measurement may be undertaken using a conductivity meter or refractometer with suitable psychrometric calibration. Soil salinity may also be conveniently determined using a flame photometer or gravimetric determination of total soluble salts. Both these procedures need calibration with a method measuring osmotic potential directly.

An interesting arid-zone phenomenon develops if even slightly saline groundwater reaches the soil surface. Evaporation will give rise to a crust of crystals and near-saturated soil solutions. A ψ_s gradient is thus set up, leading to an intensification of the ground-water flow. This process of salinisation, often confined to depressions, may lead to patterns of soil salinity and hence vegetation pattern. A secondary effect of high salt concentrations is the collapse of any clay aggregates dependent on divalent cation bridges. This process of salt-induced deflocculation leads in turn to poor drainage and poor aeration. Whether or not anaerobic conditions develop will depend on the presence of organic matter. This may well be an irrelevance where vascular plant ecology is concerned.

9

Soil atmosphere and soil temperature

SOIL ATMOSPHERE

Plant roots, soil microorganisms and fauna are consumers of oxygen and producers of carbon dioxide. This gives rise under ordinary conditions to the well known carbon dioxide concentration gradient from about 50 000 ppm (5%) in soil pores to a little over 300 ppm in the general atmosphere. The actual concentrations of oxygen and carbon dioxide in the soil atmosphere are the result of the rates of absorption or production by soil 'surfaces', the steepness of the diffusion gradient, the rate of diffusion and the length of the diffusion pathway.

Two variables are of particular importance: (a) the rates of absorption/production, which are effectively metabolic rates and thus related strongly to temperature; and (b) the rate of diffusion, which is greatly affected by the proportion of the pathway which is through water. Gases diffuse through water 10^4 times more slowly than through air. Coincidentally, and of less importance, is the increase in length of the air-filled diffusion path.

Imagine the gas diffusion pathway in a medium loam soil. There are pathways of two differing size ranges, the inter-aggregate spaces and the spaces within individual aggregates. There will thus be two trends for oxygen concentration in such a soil: a general trend towards low oxygen concentrations in the deeper parts of the rooting zone, which is accompanied by a micro-distribution pattern through each aggregate. The plant and its mycorrhizas may experience a heterogeneous oxygen concentration environment.

It is important to realise that the gas concentrations discussed are the driving variables for a more complex set of molecular relationships described by the redox potential. Thus, as an aggregated soil with rooted plants becomes progressively wetter, the centres of aggregates lose oxygen to metabolism faster than diffusion will replace it. Zero oxygen may eventually be sustained. The redox potential will also fall from positive values in the order of + 700 mV to perhaps + 200 to − 300 mV. From Table 9.1 it will be seen that, at this order of redox value, nitrate will be replaced by ammonia, and ferrous (Fe^{2+}) iron and manganous (Mn^{2+}) manganese ions will predominate over the oxidised forms. (On returning to better aerated conditions, the soluble manganese salts commonly precipitate to form nodules of black manganese dioxide, MnO_2. These are an indicator of the seasonal low redox condition.) As wetting proceeds further, such changes will extend, affecting the soil mass progressively from the base of the rooting zone. Soils in which seasonal oscillations of redox potential of this amplitude occur are termed 'gley' soils and the overall process is called 'gleisation'.

A number of consequences may result from such an episode:

(a) A combination of oxygen starvation and the toxicity of reduced ions may kill fine roots or mycelium.
(b) In the immediate proximity of larger roots, oxygen diffusion from the plant itself may provide an aerated microhabitat. The anatomy of species is important, as is its plastic response in producing more aerenchyma under progressive stress.
(c) The plant may also simply tolerate the stress, coping adequately by means of anaerobic metabolism (Crawford 1978).

As one moves progressively from aggregated terrestrial soils to those in which soil particle size, organic content or the presence of sodium restricts aggregation, or where soils are flooded by groundwater, a more extreme case develops. The microbial metabolism and/or flora may alter, leading to progressive chemical reduction of the soil system to a point where hydrogen sulphide and methane are produced. This reaction series is dependent on the presence of suitable organic substrates for the anaerobic heterotrophs to utilise. The concentration of sulphate in the substrate will determine the extent to which sulphide will be found: hence the blackening of saltmarsh sediments, as sulphate in seawater is reduced and black iron sulphides form. Stress enhancement by the reduction of metal ions, $Fe^{3+} \rightarrow Fe^{2+}$ and $Mn^{3+} \rightarrow Mn^{2+}$, will also depend on the concentration of the reducible substrate.

In all these cases there are two principal dimensions to the stress encountered by plants, the *intensity* of oxygen deficit or redox

Table 9.1: Succession of events occurring in a waterlogged soil as related to the redox potential

Period of incubation	Stage of reduction	System	Redox potential	Nature of microbial metabolism	Formation of organic acids
Normal range	Oxidised	Disappearance of O_2	+ 600 to + 400	Aerobes	None
Early	First stage	Disappearance of NO_3^-	+ 500 to + 300	Faculative anaerobes	Some accumulation after addition of organic matter
		Formation of Mn^{2+}	+ 400 to + 200		
		Formation of Fe^{2+}	+ 300 to + 100		
Later	Second stage	Formation of S_2	0 to − 150	Obligate anaerobes	Rapid accumulation
		Formation of H_2	− 150 to − 220		Rapid decrease
		Formation of CH_4	− 150 to − 220		

potential decline, and the *duration* of each episode. Approaches to measurement must take these into account, together with the idea of spatial heterogeneity mentioned above.

Several sampling and analytical procedures may be employed. These are:

(a) Direct measurement of soil atmospheric composition, now facilitated by gas chromatography.
(b) Field measurement of redox potential using a platinum electrode. Placing of the electrode is important, and it is necessary to understand the variation characteristic of a particular site before making comparisons. It seems likely that the relationship between aggregate heterogeneity and electrode dimensions is a problem.
(c) Field measurement of oxygen diffusion rate. In this case an oxygen-consuming platinum electrode is used in polarographic mode (Poel 1960; McIntyre 1970). This procedure has been used in an automated form.
(d) Field sampling and laboratory analysis of interstitial soil solution for the determination of critical ions. This is a laborious procedure and not to be undertaken lightly. Solutions are difficult to sample, transport and store and to analyse meaningfully (Martin 1968). Estuarine chemists have overcome some problems using sealed corers with sampling ports and inert gas

atmospheres. Rapid analysis for ionic species can now be achieved using anodic stripping voltametry, and this technique has been applied to sediments.

For the field ecologist, it is important to determine periods of interest for intense examination, using several techniques in complementary mode. Once a field situation is adequately described, a relationship between soil water content and a particular chemical characteristic may adequately predict duration.

SOIL TEMPERATURE

All soil activities from the decomposition of crystal lattices to the growth of microorganisms or germination of seeds are directly affected by temperature of the rooting zone. Temperature of this zone is the result of heat gain from solar radiation (1) minus reflection (3) and possibly heat flow from deeper storage (2), minus heat lost by radiation (4), convection (5), conduction (2), and evaporative cooling (6) (Figure 9.1). From the diagram it may be deduced that

Figure 9.1: Energy exchanges that contribute to soil temperature

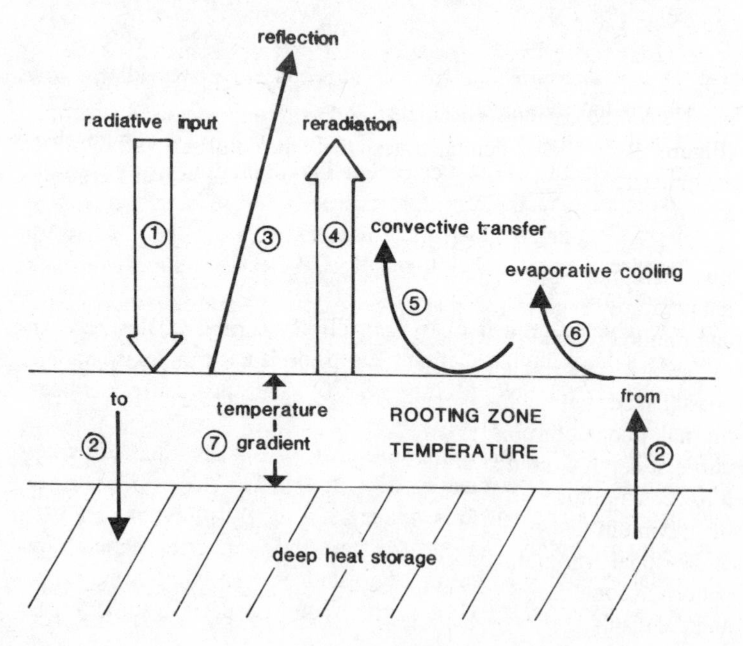

although each of the characteristics identified may be measured, and predictive models made, in practice, direct measurement of soil temperature is the main method of approach. For measurements to be of real use they should at least resolve the diurnal characteristics of the temperature gradient (7) at cardinal times in the seasonal cycle. Although heat storage and heat flow may be of great interest to the micrometeorologist, e.g. Gates (1962) and Geiger (1965), it is temperature alone that influences biological activity. The length of the growing season, the rate of processes and the cueing of processes such as seed germination are linked to temperature and duration of temperature regime.

Effects of soil components

In addition to heat inputs and outputs, a factor with a large effect on the temperature behaviour of soils is their thermal conductivity. The thermal conductivities of soil constituents vary considerably (Table 9.2). The replacement of air by water over the range of variation encountered in soils may raise the thermal capacity by a factor of three to four times, but the thermal conductivity may change by 100 times. Soils are non-uniform with depth for all of the constituents shown in Table 9.2.

In order to illustrate the likely behaviour of soil temperature, a data set for diurnal temperature variation in a loam (Yakuwa 1945) has been used as the basis for a schematic and idealised diagram (Figure 9.2). This demonstrates that the diurnal variation is asymmetric, giving a 'saw tooth' rhythm with the maxima tending to drift higher or lower on a daily basis according to season. The exact shape of the curve will depend on the various compounds of heat flux mentioned above. It is seen that with depth the amplitude of temperature range is reduced and the time of temperature maxima and minima is delayed with respect to the surface. This delay is due to the rate of thermal conduction and hence the soil composition. A coastal sand, with high conduction (see Table 9.2), will have a high diurnal fluctuation and less time displacement of the temperature curve at depth from that at the surface. In contrast, a peat will have a lower amplitude of variation and a substantial time displacement of maximum and minimum temperatures at depth.

Seasonal variations in soil temperature resemble the diurnal pattern in some respects, especially the lower amplitude of variation and the displacement of maxima and minima. In the temperate zones

Table 9.2: Thermal conductivities of soil constituents at 10°C (Hillel 1980)

Mineral	Thermal conductivity (W m^{-1} K^{-1})
Quartz	8.8
Other minerals (average)	2.9
Organic matter	0.25
Water	0.57
Air	0.025
Ice (0°C)	2.2
Snow	0.15–1.7

Figure 9.2: Diurnal temperature variation in a loam soil for two summer days in the temperate zone

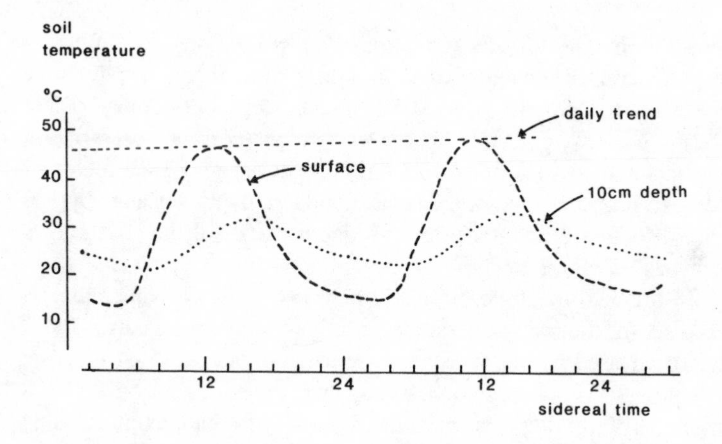

mean surface soil temperature tends to exceed mean air temperature for much of the year. The greatest differences are in midsummer.

The freezing of water in soils gives rise to a whole series of phenomena which deserve brief mention. There is about 9% increase in volume when water freezes, giving rise to mechanical pressure within the soil. The precise effect of this on structure will depend on factors such as the speed of cooling, the distribution of soil water, the soil constituents and the nature of freeze–thaw cycles. Prolonged slow cooling will tend to cause aggregates to form, separated by relatively large wedge-shaped ice crystals. Rapid and repeated freeze–thaw cycles tend to shatter aggregates with many small crystals. Stones in the soil, with water present beneath them, tend to be literally 'jacked' up through the soil. The high thermal conductivity of solids leads to rapid heat loss at night and steeper thermal gradients in the underlying water. Similar 'jacking' effects are stated to be responsible for the rupturing of seedling roots (Hillel 1980).

The phenomenon of permafrost, a permanently frozen lower soil horizon with variable depth of summer thaw, is the ultimate low temperature effect and characterises the tundra biome. The complex web of environmental and biological interactions is dealt with in Chapter 17.

10

Some examples of mineral nutrient supply

INTRODUCTION

The examples of mineral nutrients chosen for discussion are nitrogen, phosphate, potassium, magnesium and copper. The basis for choice is a combination of absolute biological and ecological importance plus the capacity for illustrating more widely applicable phenomena. Nitrogen and phosphorus are the two elements most frequently serving as limiting factors in plant growth, nitrogen being uniquely complex with many biological processes and several chemical forms in its cycle, whereas phosphorus is complex mainly from a geochemical viewpoint.

Potassium is a biologically important univalent cation, and magnesium represents a divalent example. Both elements can occur at high concentration, with magnesium being occasionally toxic. Copper is an example of an essential trace element with a continuum of effects from deficiency through normal sufficiency to acute toxicity.

This account will concentrate on the environmental and biological processes that directly affect plant growth, and will do so in the context of the general cycling pattern for the element in question. The most difficult environmental variables to measure are those concerned with the supply of ions. This chapter and Chapter 11 are concerned with the 'why' and the 'how' of approaches to these problems.

SYNOPSIS OF THE NITROGEN CYCLE

The complex and unusual cycle for nitrogen (see Figure 10.1) has at its core the familiar pattern of absorption of inorganic ions from soil, transfer through ecosystem processes, mineralisation and re-entry to the liquid phase as ions. The outstanding feature of the cycle core is that virtually the only currency of nitrogen transfer through the ecosystem is protein. Plants are quantitatively the major synthesisers of protein from ionic nitrogen followed a long way behind by microorganisms. Protein nitrogen abounds in ecosystems, ionic or crystalline forms being minor and transient.

The most important difference between the behaviour of nitrogen and other mineral nutrients is the extent of exchange with the environment at large, both output and input. The two processes contributing to output are (a) the leaching of the nitrate (NO_3^-) anion from soil, soil virtually lacking anion exchange sites; and (b) denitrification, the microbially mediated process by which nitrate is reduced to the gases nitrous oxide (N_2O) and dinitrogen (N_2). In some ecosystems, burning of vegetation is an additional regular pathway for nitrogen removal.

Input of nitrogen occurs mainly from the biological reduction of dinitrogen to amino groups and ultimately protein, that is, biological nitrogen fixation. The atmospheric pool of dinitrogen, and soil atmosphere in particular, is an important segment in the cycle. Air pollution may also contribute substantial annual inputs of fixed nitrogen to some ecosystems, with quantities in the order of 10–30 kg of nitrogen per hectare being common in Europe. The particular problem of supply of nitrogen to plants in ecosystems is linked to this web of interactions, all of which are affected by soil conditions in terms of reaction rate and relative importance.

PHOSPHATE

In most ecosystems the supply of phosphate to a square metre of vegetation depends almost entirely on the phosphate content of the square metre of substrate beneath. Why should it be necessary to make such a simplistic assertion? First, it is patently not true of elements such as nitrogen, where biological fixation and leaching are important in the nutrient cycle; or potassium, where inputs through rainfall can be of great importance. Secondly, this idea, I believe, is useful in the interpretation of ecological situations. It is suggested that a crude but helpful first approximation of soil fertility may be derived from total phosphate concentration (Table 10.1). A suitable starting point for describing the supply of phosphate to

Figure 10.1: Synopsis of the nitrogen cycle. See Commentary for explanation

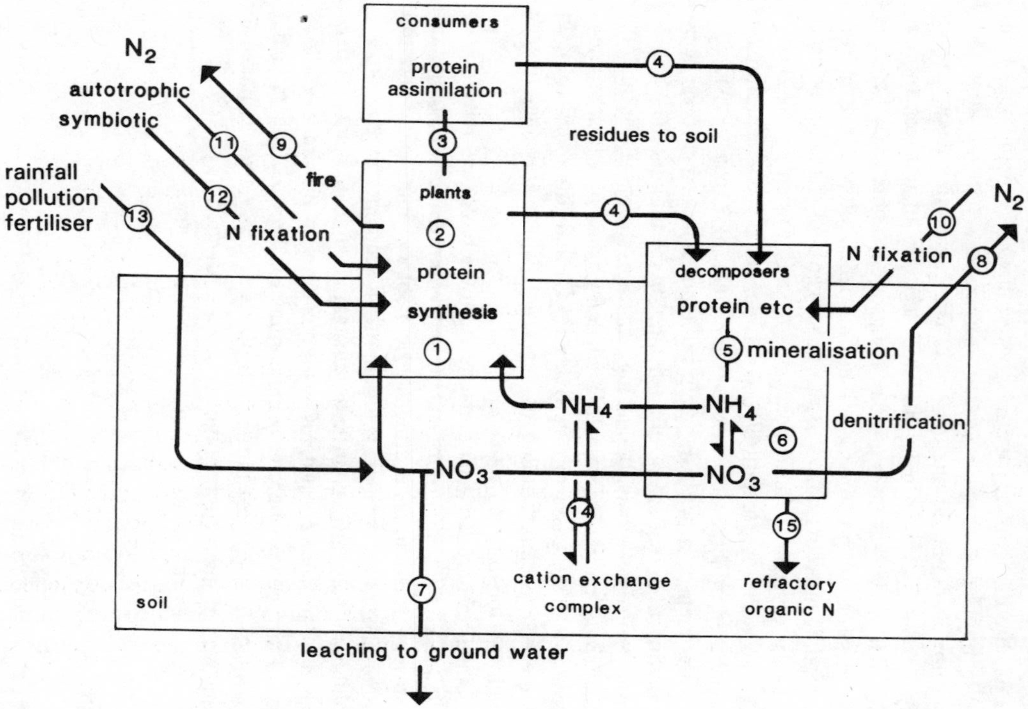

Commentary on Figure 10.1

① Nitrogen assimilation is strongly linked to the growth of the plant, the activity of the root system and hence to climatic conditions. See also Figure 2.6 for enzyme complement.

② Protein synthesis is dependent on carbon skeletons (photosynthesis) and overall growth pattern of plant (metabolic sinks).

③ Predation requires the digestion of plant protein and resynthesis of characteristic molecules by the predator. Because some protein is catabolised, nitrogenous excretion occurs.

④ Residues transferred to soil include a wide range of molecules, from protein and chitin to uric acid, urea and ammonia.

⑤ Ammonification: this should be seen as a series of reactions, with microbial types utilising various molecules from urea to nucleic acids. Rate is dependent on environmental conditions including temperature, pH, water potential and aeration. C:N ratio is also important as this is a net process between release and assimilation by the changing microbial population. Ammonification is not inhibited by $pH < 4.0$. In calcareous soils ($pH > 7.5$) ammonia may be lost in significant quantities by volatilisation.

⑥ Nitrification from ammonia is a process accomplished by mixed microbial populations, with nitrite seldom detectable. Rate is sensitive to the above environmental factors but is comparatively more sensitive to pH. Rate drops rapidly below pH 6 and is negligible at pH 5. Adequate aeration is necessary. Specific inhibitors of nitrification have been detected in the field.

⑦ Nitrate leaching can only occur when nitrate is being produced in the absence of assimilation. This can occur in cold soils in the temperate-zone winter period. Rate is a function of rainfall and soil permeability.

Commentary on Figure 10.1 *Continued*

(8) Denitrification: this may be a major source of nitrogen loss if soil conditions oscillate between aerobic and anaerobic. Denitrifying bacteria are all aerobic but use nitrate as the electron acceptor in the absence of oxygen. Another key environmental factor is pH, with the process being of significance only at pH > 5.0. The reaction produces a mixture of dinitrogen (N_2) and nitrous oxide (N_2O).

(9) Fire must be included as a catastrophic, but recurrent, process in some ecosystems that releases organic nitrogen capital as dinitrogen and oxides of nitrogen. It also leads to chemical mineralisation.

(10) Heterotrophic nitrogen fixation is dependent on energy availability in the form of organic matter. The most effective systems may be associated with the plant rhizosphere.

(11) Autotrophic nitrogen fixation is represented mainly by cyanobacterial masses or films at the soil surface. Nitrogen fixation activity may be transient, controlled especially by soil moisture.

(12) Symbiotic nitrogen fixation involving vascular plants is covered in Chapter 4, but cyanobacterial lichens may also contribute. All nitrogen fixation is inhibited by soil nitrate. Optimal activity requires a pH close to 7.0, and relatively high phosphate, calcium and micronutrient status.

(13) Miscellaneous inputs. The most significant of these are nitrogen fixed by lightning discharge, pollution from high temperature combustion, and industrial fertiliser production.

(14) Equilibration of NH_4^+ with cation exchange sites, a source of available nitrogen.

(15) Refractory organic nitrogen which includes chitin and proteins 'tanned' by reaction with soil phenolics.

Table 10.1: A first approximation towards interpreting soil total phosphate content. Data from many sources, corrected for soil bulk density

Total soil phosphate content (mg P per litre of soil)	Suggested vegetation
Less than 50	Heaths and bogs. Rarely encountered (Chapter 17) and regarded as an extreme condition
50–75	A transitional situation with species-poor grasslands and scrub of low productivity. Some forest types may establish at the upper limit e.g. *Pinus sylvestris, Betula pubescens, Pinus radiata*
75–250	Very commonly encountered in temperate soils, and phosphate is probably not a limiting factor. This includes most natural and plagioclimax grasslands, sand dunes, saltmarshes and many woodland soils. Phosphate mineralisation and availability are of great importance
250–500	The most productive natural woodland soils, now mainly under agricultural production
Greater than 500	Seldom encountered without fertiliser addition. Temperate rainforest

ecosystems through plants is to enumerate the solid-phase sources of phosphate (Table 10.2).

In Figure 10.2 it is seen that the outcome of the behaviour of the inorganic solid-phase phosphates results in iron and aluminium phosphates predominating at low pH whereas only calcium phosphates of the apatite family exist in high pH soils. (This state of affairs is reflected in several commonplace matters including the inclusion of phosphoric acid in rust inhibitors, the preservation of iron objects in acid soils, the vulnerability of dental enamel to organic acids, and the protective action of fluoride in the apatite crystal lattice.) A prediction of the fate of added soluble phosphate, fertiliser or excreta for example, is also evident from the model. Thus rock phosphate, an impure apatite, is a suitable fertiliser for

Figure 10.2: Solubility of solid-phase phosphate minerals with pH. The *y*-axis represents the concentration of phosphate ions in equilibrium with each solid-phase source

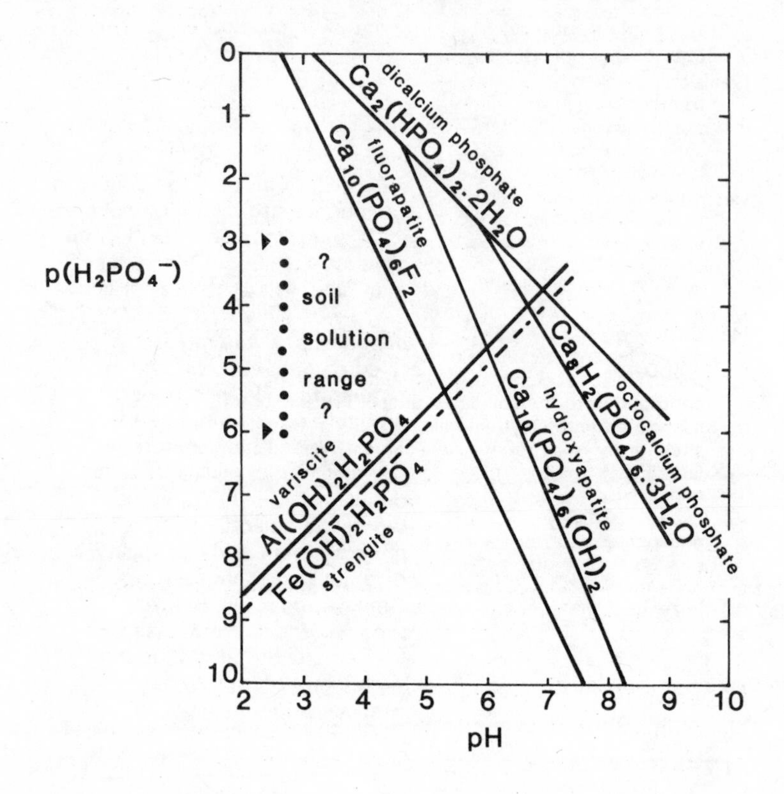

forest trees on acid soils, but will be immobilised eventually as iron or aluminium phosphates.

Organic phosphate

Although literature reports indicate the detection of many biochemical phosphate esters in soil extracts, including phospholipids, sugar phosphate and nucleic acids, the quantities are small and their origin and ecological significance are uncertain. It is currently accepted that the major group of organic phosphates are the inositol phosphates and in particular the isomers characteristic of microbial metabolism rather than residues from higher plants

Table 10.2: Solid phase sources of phosphate in soil

(1) *Apatite minerals*, for example hydroxyapatite, $Ca_{10}(PO_4)_6(OH)_2$. Many substitutions are possible, and apatites are found as primary minerals and secondary minerals dependent on soil development. Most important are the substitution of OH by F (see Figure 10.2), CO_3 for PO_4 and several divalent metals for calcium, e.g. lead.

(2) *Other calcium phosphates*. Agricultural studies show that a number of other calcium phosphates, e.g. octocalcium phosphate, may be present in soils. These may be regarded as transient types at low solid-phase concentration. They may make a disproportionately large contribution to the soil liquid phase, as 'labile phosphate'.

(3) *Iron and aluminium phosphates*. These are weathering products rather than primary minerals, forming as microcrystalline deposits associated with oxide and hydroxide surfaces of soil. The iron salt, strengite ($Fe(OH)_2H_2PO_4$) releases phosphates when potential is lowered. Iron–aluminium–calcium phosphate ratios have been used as a soil weathering index.

(4) *Other inorganic phosphates*. In highly weathered tropical soils, phosphate minerals may be coated with deposited iron or aluminium salts. These 'occluded' phosphates obviously make a minimal contribution to the soil liquid phase. Phosphate ions are also associated with clay minerals, loosely in the few anion exchange sites, or by isomorphous replacement with silicate as part of a lattice. Contribution to the liquid phase may be substantial in the middle of the working range of soil pH in agricultural soils.

(5) *Organic phosphate*. Predominant are various forms of inositol phosphate, as calcium or iron salts. The isomeric forms of the inositol skeleton indicate a microbial origin. Sugar phosphates, phospholipids and nucleic acids are also present in soil extracts, but are probably short lived in mineralisation terms.

(Cosgrave 1980). Generally the proportion of organic phosphates is higher in soils with low total phosphate. This makes this source of greater interest to the ecologist than to the agriculturalist.

Release of phosphate to the liquid phase

The picture gained of the phosphate radical so far is of a reactive trivalent anion forming rather stable compounds with tri- and divalent metal ions and extremely stable esters with the cyclic sugar inositol. It is also intrinsically at a lower concentration in the fabric of soil than calcium or potassium, for example. Because of these properties, phosphate diffuses very slowly in a soil matrix and is very slowly removed from soil by leaching. One estimate of the half-life of the phosphate in soils developing on volcanic ash in the warm temperate climate of New Zealand is 20 000 years. Nevertheless, phosphate is absorbed by plants and is detectable in the liquid phase of soil. What are the factors influencing a shift in the equilibrium from solid to liquid phase?

In the case of crystalline inorganic compounds we must anticipate a solubility-product-type equilibrium, but it would be naive to expect pure water as the solvent. It is particularly difficult to handle the 18-ion apatite series, even when both lattice composition and bathing solution are defined (Larsen 1967). This means we must resort to a virtually empirical approach to understanding the equilibrium, solid phosphate → liquid phosphate. It will be shifted from left to right if the following occur:

(a) Removal of phosphate by absorption, the zero sink created at the surface of an organism.
(b) pH shift affecting solubility of iron, aluminium or calcium phosphates.
(c) Chelating agents, liganding these metals, encourage solution and prevent reprecipitation.
(d) Reduction in redox potential leading in particular to change from Fe^{3+} to Fe^{2+} and ultimately FeS precipitation. Iron phosphates may massively dissolve under these conditions, a state of affairs encouraged in the flooding of rice fields.

All four of the above may be imagined to occur as a result of root, mycorrhizal or microbial metabolism.

Organic phosphate compounds in soil, including the inositol phosphates, can be imagined to supply phosphate ions to the liquid phase only when enzymically hydrolysed. Several workers have shown that root surfaces, including their rhizosphere and mycorrhizas, possess phosphatase activity. This assay, using paranitrophenol phosphate (PNPP) substrate, is a comparatively simple procedure that might be used more frequently in ecological situations.

paranitrophenylphosphate paranitrophenol
PNPP + phosphatase → PNP + phosphate ion
(colourless) (yellow)

The knowledge of the reactivity of the phosphate ion and its resultant slow diffusion in soil emphasises the need for the root system to explore soil volume. Optimisation of the absorbing surface to soil volume ratios is probably approached by mycorrhizal roots, with extracellular phosphatases also making a contribution to absorbable phosphate (see Chapter 4).

Another feature of phosphate supply is that it seems to vary from season to season, as if it is generated in 'pulses'. This is well marked at low concentration. Freeze–thaw cycles are stated to be a cause of phosphate release from organic soils to *Rubus chamaemorus* roots

(Saebo 1969); and Rorison (1969) suggests that an irregular supply is assimilated by *Rumex acetosa* on calcareous soils. Even agricultural studies (Williams 1955) illustrate that a great deal of phosphate absorption takes place at the very beginning of the growing season of an annual crop such as oats. Fertiliser application experiments on forest crops show that if phosphate is applied later in the season, 'spring' wood, with long fibres, continues to be produced. This implies that here a spring flush of nutrient is essentially responsible for the rapid growth increment. It should be noted that nitrogen supply is also episodic, and phosphate and nitrogen pulses may coincide when they are linked to biological mineralisation phenomena. This is a field well worth studying further, as it may be seen that phenological changes are linked with nutrient pulses.

Alternatively, phosphate may be stored after absorption until conditions for growth are favourable (see Chapter 16).

POTASSIUM

Potassium (Figure 10.3) is an abundant element in the Earth's crust, soils, plants and animals. Supply does not act as a limiting factor to growth in uncultivated systems. In agriculture it is certainly of great importance, and most of the information stems from applied studies with fertilisers.

Primary minerals

Potassium is associated with the most abundant minerals in the Earth's crust, the feldspars. Weathering of feldspars to yield potassium ions has already been described as an example of weathering in soil formation (Chapter 7). Another abundant source is the layer silicates or micas, especially biotite and muscovite. The process is slow and cannot match the demand by vegetation.

Exchangeable potassium

The cation exchange complex is the only source of rapidly available potassium to an organism. In a natural ecosystem it is assumed that depletion of the cation exchange complex is replenished by mineralisation. This is usually rapid, as potassium is readily leached from dead tissues and fairly easily removed from living leaves (see Chapter 15).

The pool of exchangeable potassium may also be replenished by

Figure 10.3: Comparison of potassium and magnesium supply to the soil liquid phase

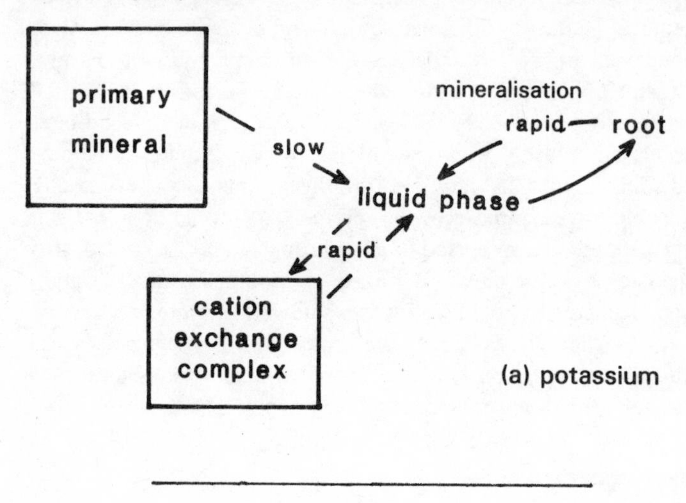

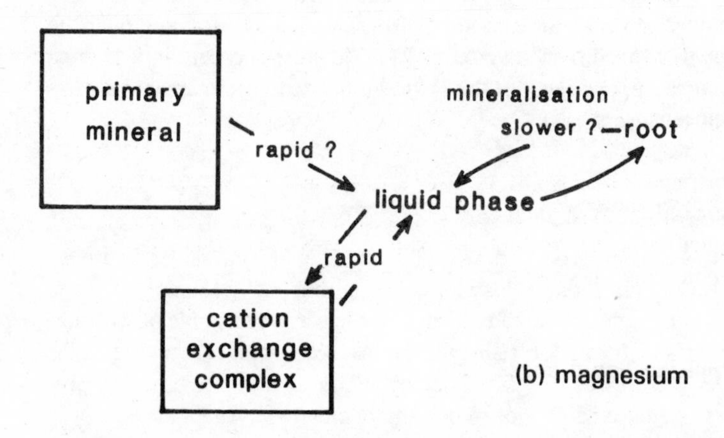

inputs from rainfall or dust. Leaching is clearly a mode of potassium removal. The relatively rare element caesium (Cs) is very similar to potassium in its behaviour in soil. Thus this is the model to anticipate the behaviour of the radioactive isotopes ^{137}Cs and ^{134}Cs after their dispersal by the reactor explosion at Chernobyl in the Ukraine in spring 1986. Since the half-life of ^{137}Cs is 30 years, there will be time to explore many aspects of absorption and turnover of a potassium-like element in many European ecosystems. Sensitive low-level isotope counting techniques, including gamma-spectroscopy, can characterise close to background activity levels if a vegetation sample in the order of 1 kg dry weight is available.

MAGNESIUM

Magnesium (Figure 10.3) is another abundant element in both the crust of the Earth and in the layer below the crust, the mantle. The mantle became a resting place for the magnesium–iron silicates which melt less easily than the feldspars and are more dense. The principal minerals in this zone are olivine ($Mg_2SiO_4 . Fe_2SiO_4$) and pyroxine ($MgSiO_3 . FeSiO_3$). These occasionally give rise to soils, for example on basalt flows.

Primary crustal minerals

In the crust, as distinct from the mantle, many common aluminosilicates include magnesium atoms along with calcium and iron. These minerals account for the majority of magnesium sources in most soils. Another common situation is the presence of magnesium carbonate in the so-called 'dolomitic' limestones.

A number of rare mineral types give rise to high-magnesium substrates. These are the very soft magnesium silicates, serpentinite, $Mg_6Si_4O_{10}(OH)_8$, and talc, $Mg_3Si_4O_{10}(OH)_2$. The serpentine anomalies of the world give rise to an enigmatic ecological phenomenon which is examined in Chapter 19.

Supply to liquid phase

Because of the variety of geochemical situations, magnesium supply is likely to be more complex than that of potassium. Undoubtedly magnesium ions are held by the cation exchange complex in equilibrium with soil solution. However, in some cases — dolomitic limestone, for example — direct solubility may yield a source matching the demands of growth.

COPPER

There is a great deal known and much to be learned about this trace element (Nriagau 1979; Loneragan, Robson and Graham 1981). The problem is that geochemists and inorganic chemists can give a lengthy account valid only for the theoretical solid-phase minerals and complex ions in relatively concentrated systems. In soils, however, we are normally dealing with low concentrations, 70 ppm in the Earth's crust giving rise to about 20 ppm total copper in soils. The predominant question is what controls the copper ions that are absorbed by organisms. These will be assumed to be hydrated divalent ions $Cu(H_2O)_6^{2+}$ in soils of pH 5–8 (Parker 1981).

Primary minerals

Although there are a large number of possible primary minerals, sulphides, oxides, carbonates, silicates and chlorides, measured concentrations of copper ions in soil solution are too low to be due to a direct solubility–product relationship (McBride 1981). Copper sulphide, added experimentally to plant growth media, is an effective depressant of growth (see Figure 14.3).

Adsorbed copper

Copper ions are strongly adsorbed to the surfaces of both silicate clays and oxide clays (see Table 8.3). Organic matter is another very effective adsorber and may well contribute to most exchangeable copper ions.

Chelating agents such as EDTA remove adsorbed copper with great efficiency and may be used in the estimation of available copper. It is interesting that the concept of availability may be applied equally to the system fulfilling the need for an essential trace element and to the potential of a substrate for toxicity.

Ligand-bound copper

Organic substances in soil may also bind copper through the activity of several groups but especially COO^-. Solid-phase and liquid-phase ligand formation is likely. The former is thought to be responsible for both the induction of deficiency symptoms in organic soils and the relief of toxicity symptoms by addition of peat and other organic compounds to high-copper substrates (Jeffrey and Maybury 1981) (see Chapter 14). Liquid-phase ligands, probably of low molecular weight, facilitate mobility of copper down profiles and in moving groundwater.

11

Measuring availability of nutrients and toxic ions

SOLID PHASE–LIQUID PHASE RELATIONSHIPS

It was taken for granted earlier (Chapter 1) that the soil–plant–air continuum is a study unit of value in discussing transfers of water, minerals and energy. It is now worth focusing on the first part of the continuum in the light of the processes of soil formation, but with a view to understanding nutrient flow to plants and ecosystems.

The solid-phase fabric of soil is a mixture of rock-fragment-derived inorganic particles plus organic material derived from primary or secondary production. The liquid phase is soil water. As an example, imagine a totally non-living system comprising feldspars from granite, organic matter and water. The potassium aluminosilicate mineral will react with water, especially if excess hydrogen ions are present, to give the clay mineral kaolinite. This is basically the reaction giving rise to china clay (Figure 8.4). It may be represented as:

potassium feldspar		kaolinite		silicate and potassium ions

$$2\ KAlSiO_8 + 2H^+\ H_2O \rightarrow Al_2Si_2O_5(OH)_4 + 4SiO_2 + 2K^+$$

Some of the silicon and all of the potassium appear in solution. Both organic matter and the kaolinite have cation binding sites which may remove K^+ ions from solution in exchange for H^+ or other bound cations including Ca^{2+} or Mg^{2+}. Depending on particle size, between 10 and 50 mg of potassium per litre may be present in such a system at equilibrium. This example illustrates the concept of solid phase–liquid phase interactions involving the alteration of minerals and the exchange of ions, emphasising the totally inorganic transactions.

Now add a living sink for potassium ions, a root, which will tend to reduce the concentration in solution. As time passes, the potassium ion concentration in solution will be replenished by two sources, (a) the exchange of K^+ ions from organic matter and kaolinite surfaces for whatever counterion is released by the absorbing root; and (b) the continued breakdown of K-feldspar, which is a very slow process in relation to cation exchange.

Perhaps this is not the most simple example of a solid phase–liquid phase situation, but it illustrates the relationships between rock decomposition and the properties of secondary minerals and their role, together with organic matter, as a solid-phase ion reservoir. It also makes clear the assumption that ion uptake by organisms is an equilibrium-shifting process that may feed back to rock weathering, if only through release of H^+ ions.

In the course of discussing supply of individual elements more examples will be dealt with which range from the dissolution of calcium carbonate to the complex biological and chemical process steps that release nitrate and ammonium ions to the liquid phase.

AVAILABILITY

The conceptual complexity of ion transfers, the physical difficulties of gaining experimental access to the soil–root interface, and the need to integrate data over very long or very short time periods are all obstacles to a useful working knowledge of these systems. Agriculture and ecology have to resort to a mixture of theory and empiricism in qualifying actual and potential flow of ions between soils and plants. This is summed up by the term 'availability'. Because it is used, and misused, by the two disciplines, it is worth making an effort to clarify its meaning and review its usefulness.

In agriculture, the concept of availability concerns the capacity of soil to deliver to the crop a supply of essential ions. This may be redefined in terms of solid phase–liquid phase plant relationships as the extent to which removal of ions by the plant is restored by the solid phase. A major economic requirement for agricultural management is to determine the return, in crop growth terms, on a given investment in fertiliser. In agriculture the prime need is to find an easily determined soil characteristic that correlates well with crop yield to serve as a useful index of some aspect of soil fertility.

The ecologist may wish to possess information about different aspects of the system. Typical questions include:

- Which ion supplies are fundamentally limiting to biomass production?
- What is the rate of ion supply to an ecosystem?
- How does the soil-fertility component of the environment vary from site to site?
- Have given species particular soil-fertility requirements?
- How accessible to plants are potentially toxic elements in soil?

With the growth of thinking about agro-ecosystems and the realisation that toxicity may occasionally be important in agriculture, e.g. selenium toxicity, there has been some convergence between the two fields.

Practical approaches to availability

Several different approaches to availability are found in common use in agriculture. It helps to divide them into categories before adapting them for use in ecology.

(1) Determination of a clearly defined soil characteristic, e.g. exchangeable + soluble soil potassium as extracted with ammonium acetate. This is the best short- and medium-term measure of potassium supply, and relates directly to the presumed soil supply system. Hot-water-soluble borate is a simple effective measure of boron availability.
(2) Extraction or displacement of soil liquid phase, e.g. estimation of NO_3 in soil. May be repeated to determine nitrification rate.
(3) Utilisation of natural ion supply system, e.g. exchangeable potassium (above). Estimation of ammonium and nitrate supply through mineralisation by incubation methods. The natural microbial population is incubated to accelerate release of the two ions.
(4) Extraction using chemical reagents empirically selected because of high correlation with crop response or element in crop. Some of these have been designed around fundamental soil characteristics, but it is emphasised that they are seldom conceived of as mimicking or modelling the supply system in the soil. Favoured because of low cost and speed. For example, ammonium bisulphate at pH 3 (Truog's reagent) for available phosphate: there are a very large number of phosphate

availability systems of this kind, designed on the whole for use in closely defined agricultural situations. Most need re-evaluation before being applied to ecological situations. Another example of such an extraction method is the use of EDTA for toxic heavy metals.

(5) Use of a living plant, e.g. Neubauer test for phosphate in which nutrients removed from 100 g soil by 100 rye seedlings are determined. These systems are often able to rank soils but quantitatively depend on individual species–soil relationships. Too slow for most routine purposes.

(6) Use of an artificial 'sink' as a replacement for the plant, e.g. the 'resin' method for phosphate which entails adding an anion exchange resin to the soil system and determining phosphate ions absorbed. This may turn out to be a good method for ecological use as the resin mimics the 'zero sink' of the root or mycorrhiza.

(7) Extrapolation from measures of total elemental composition. These may give a useful first approximation for ranking soils, especially when results are expressed in soil volume terms. This approach is of great value in detecting anomalous base metal concentrations, especially if an analytical system such as X-ray fluorescence is used.

(8) Use of nutrient addition experiments. These are employed in agriculture as a way of assessing the value of empirical availability tests, but are generally very expensive and cumbersome. In ecology they are of very great value in certain situations, and will be discussed in more detail later (Chapters 12 to 14, 16, 18 and 19).

ION EXCHANGE PROPERTIES

The most rapidly equilibrating reactions in the solid phase–liquid phase system are those involving the exchange of cations between soil solution and negatively charged surfaces. The presence of such surfaces is attributable (a) to organic matter and especially its carboxyl and phenolic hydroxyl groups; and (b) to minerals with the silicate clay minerals and hydrous oxides of prime importance.

The cation exchange capacity (CEC) of soil as a whole is conventionally expressed in terms of milliequivalents of exchange capacity per 100 g of soil. This rather awkward unit merely gives values in a convenient range between about 20 and 100 meq per 100 g. To give some order of importance to the components of cation exchange capacity, the following values in milliequivalents for each 1% contribution, were suggested by Thompson (1957).

Organic matter	2.0
Vermiculite	1.5
Montmorillonite	1.0
Illite	0.3
Kaolinite	0.1

This is not an approach to determining cation exchange capacity, but the values indicate the importance of organic matter and show that generalisations cannot be made about the clays without a mineralogical analysis.

Selectivity

Cations vary in two major ways, their valency and their size, i.e. ionic radius (see Table 8.2). Size is also dependent on hydration. Thus, relatively large and univalent cations are displaced from exchange positions more readily than smaller ions with higher valency. For the common cations the series in order of decreasing ease of replacement is:

$$Na^+ < H_3O^+ < K^+ < NH_4^+ \ll Mg^{2+} < Ca^{2+} \ll Al^{3+}$$

This means, for example, that calcium ions in the liquid phase will tend to displace, or exchange with, pairs of sodium ions bound to surfaces, assuming the same equivalent concentrations in the system. Swamping with a high concentration of a particular ion will lead to an equilibrium in favour of its being bound. Cation exchange capacity is routinely determined by treating the soil sample with a relatively concentrated solution of an ammonium salt, usually ammonium acetate. This effectively saturates all exchange sites with ammonium (NH_4^+) ions and displaces pre-existing exchangeable cations. This procedure is then utilised in two ways to determine total CEC and its make-up:

(1) After filtration and removal of any residual ammonium acetate solution, all the exchangeable ammonium is stripped from the soil with a potassium chloride solution and the ammonium ions are determined in millequivalents of ammonium per 100 g of

soil. This value is an estimate of the total cation exchange capacity of the soil.

(2) The ammonium acetate solution used to saturate exchange sites now contains all the cations originally present. The dominant cations are usually hydrogen, potassium, calcium and magnesium. In coastal soils sodium ions and in acid soils aluminium ions (Al^{3+}) may also be present. Analysis of this solution for cations other than hydrogen ions, and expression of individual values and their sum as milliequivalents per 100 g soil, permit comparison with the total cation exchange capacity.

The difference between the exchangeable bases and the total cation exchange capacity is attributable to H^+ ions. Total exchangeable bases as a percentage of total cation exchange capacity is normally referred to as 'percentage base saturation'. This is a convenient index of potential cation supply to plants. The usual ratio of bases in temperate zone soils with high base saturation is in the order of: calcium 80–85; magnesium 15–20; potassium < 1; sodium < 1. The only other ion of importance, ammonium, must be measured independently. The bases in this system are maintained in a dynamic equilibrium between the ecosystem processes of circulation and the Earth processes of weathering and leaching. In a soil system in which resistance to weathering is combined with high leaching rates, base saturation will tend to fall. In this case hydrogen ions become increasingly important, both in occupying cation exchange positions and in the soil solution. Thus by measuring hydrogen ion concentration in soil solution directly, which we can do using a hydrogen-ion-sensitive glass electrode and a suitable meter, we are indirectly estimating base saturation. For similar soil types in a region, it would be reasonable to expect a useful correlation between soil solution hydrogen ion concentration and percentage base saturation but the precise slope and intercept of such a curve would depend on the nature of the exchange complex.

Hydrogen ion should be regarded as a soil cation comparable with potassium, calcium or magnesium in cation exchange terms. The use of the logarithmic pH scale unthinkingly has tended to obscure this relationship. Current work on the input of acid precipitation to catchments certainly demands this approach. In order to convert pH values to μmoles hydrogen ions per litre, use the formula: $1/\text{antilog pH} \times 10^6$.

Role of cation exchange systems

It should now be appreciated that there are two independent sources of ions contributing to the liquid phase: hydrolysis of minerals by weathering processes; and the cation exchange complex (CEC). The contribution of organic matter to CEC is a very direct link between soil development and ecosystem activities. The whole cation exchange system serves to buffer the solution phase against depletion by either uptake by organisms or leaching. This may exert an influence on weathering rates because of the buffering of hydrogen ion concentration and weathering products.

Causes of pH change

Over a long period of time, that is, in terms of soil development, the general soil pH is a result of the size of the cation exchange complex and the extent to which it is supplied with cations other than hydrogen ions. An interesting case of a very large exchange capacity being generated is that of raised bog peats. As a bog develops from a minerotrophic basin to a dome dependent on rainwater ions, the base saturation drops whereas the total CEC remains high. To a lesser extent this phenomenon operates wherever humus is seen to be accumulating or only slowly decaying.

Acid soils can also be expected where slow-weathering rock types and a leaching climate are superimposed. Commonly this means the older mountain ranges where igneous or hard sedimentary rocks are weathered by montane or oceanic climates. Removal of bases down the profile, especially if the base : silica ratio is low in the original rock, will produce low pH soils irrespective of the size of the exchange complex. The other extreme is the soil where soluble base ions are very readily supplied by weathering to saturate the exchange system and the soil water. Coastal saline soils have pH values between pH 7.5 and 8.5, reflecting the presence of calcium carbonate and the usual pH of coastal seawater, which is close to pH 8.5.

Extremely low pH may develop in soils that contain quantities of the mineral iron pyrite, FeS. In this case both microbial and chemical pyrite oxidation leads to the generation of sulphuric acid. This case is uncommon, but may be encountered regularly in the ecological literature associated with coal-mine-waste revegetation (see Chapter 14).

EFFECTS OF SOIL pH

Soil pH is the most frequently measured and quoted characteristic in ecological and agricultural soil literature. It must be recognised that the immediate and direct effects of pH within the range pH 8.0–4.0 on plant growth are negligible, given that all other environmental features are equal. This can be stated for the relatively few cases where growth in water culture over a wide pH range has been studied. Soil microorganisms appear to exhibit some direct responses to pH.

There are many second-order effects, which contribute to a correlation between plant performance and pH, and which make interpretation of pH data alone very speculative. Although several aspects of this topic will be examined in more detail, it is worth listing a series of headings as a summary.

1. Soil fertility

The qualitative nutrient availability diagram which was produced by Truog in 1946 as a guide for agriculture is worth reproducing (Figure 11.1). It must be pointed out, however, that agriculture is concerned with managing the soil environment to optimise productivity of a few selected plant species. Ecosystems are found on soils across the pH range indicated, and even further into the acid end of the spectrum. Calcareous soils, in which calcium carbonate as rock or seashell fragments weathers to saturate the soil with divalent calcium ions, are the most commonly encountered examples. In a soil system with solid-phase calcium carbonate, an important determinant of pH is the partial pressure of carbon dioxide. The effect of carbon dioxide on pH of a pure water system, namely

$$H_2O + CO_2 \rightarrow H^+ + HCO_3^-$$

is seen to entail a reduction in pH from about pH 6.2 with atmospheric concentration of CO_2 (i.e. 300 ppm) to pH 4.5 at 5% CO_2, a value typical of soil atmospheres. (This is the phenomenon contributing to the acidity of carbonated mineral waters.) Reductions in soil pH of this order of magnitude can be instantly seen as a way in which roots and soil organisms generally affect soil chemistry. When $CaCO_3$ is introduced, the system may be represented as follows:

$$CaCO_3 + H_2O + CO_2 = Ca(HCO_3)_2 = Ca^{2+} + 2(HCO_3)^-$$

The curves associated with this system explain why the pH of a calcareous soil is seldom greater than pH 8.3 and how the substantially increased solubility of calcium carbonate under carbon dioxide

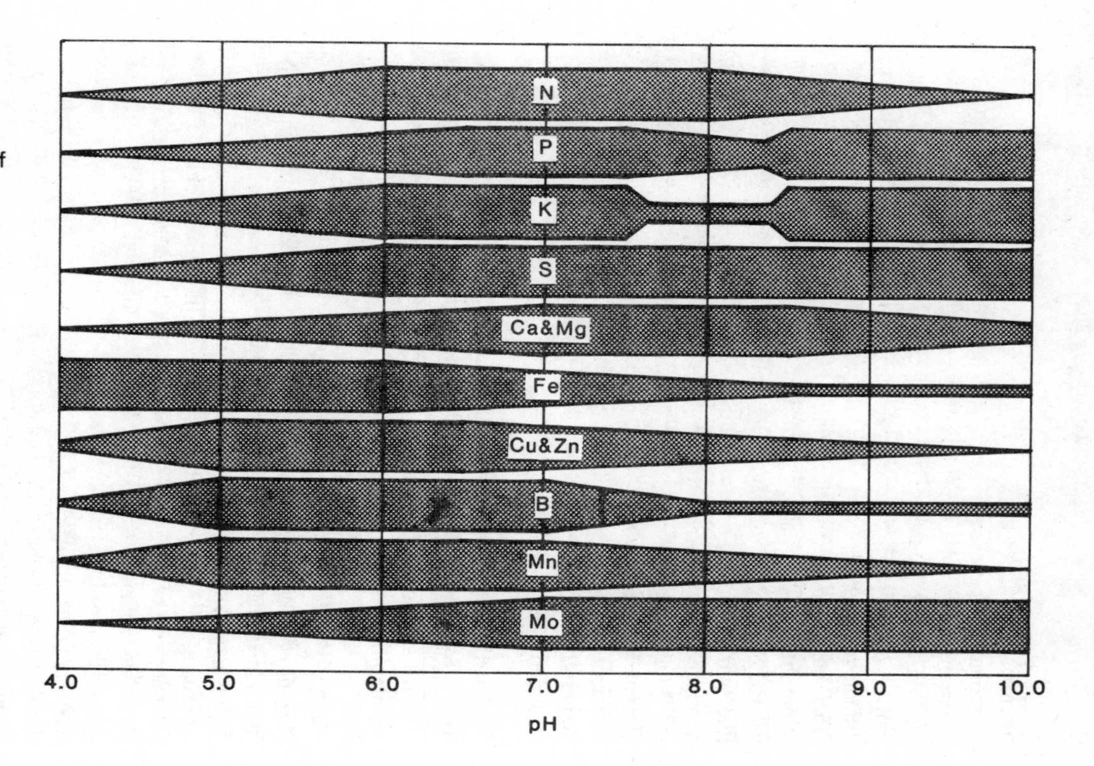

Figure 11.1: Truog's diagram illustrating the comparative availability of essential mineral nutrients as a function of soil pH

regimes encountered in soil leads to decalcification. As soil carbonate content declines, so pH falls, as shown by an empirical relationship in soils from Derbyshire (UK) (Grime 1963).

Soil alkalinity greater than that encountered in calcareous soils only occurs when sodium is present. The hydrolysis of sodium carbonate produces the very strong base NaOH. When soil is more than 15% base saturated with sodium, pH may be between 8.5 and 10. Note that a 10^{-4} N solution of NaOH in pure water will have a pH of 10. Such soils are naturally confined to the saline arid zones of the world, but high soil salinity may also develop under poorly managed irrigation.

What is interesting is the tendency for simple correlations to be absent. Truog's diagram also assumes that an optimum quantity of nutrient elements is present, but in ecological situations this is generally not the case. Low element concentrations may be confounded with availability.

2. Base status

Base status, although allied to fertility, is related directly to pH in a more simple fashion than any of the other components of nutrient availability.

3. Soil toxicity

The most widespread elemental toxicity is that of aluminium, which is toxic to many plants at concentrations greater than 1 ppm in the soil water. Aluminosilicate lattices collapse progressively at low pH, releasing Al^{3+} ions which may saturate the base exchange system. The activity of Al^{3+} ions increases markedly below pH 5 (Magistad 1925) as demonstrated for tropical soils by Figure 11.2. This may prove an absolute barrier to survival and colonisation for sensitive species (Rorison 1960). The experimental aspects of this topic will be discussed in terms of calcicole/calcifuge ecology (Chapter 19). Iron and manganese also become more available at a low soil pH but their toxicity depends on oxidation state, i.e. low redox potential (see Chapter 9). The other family of toxic metals, cadmium, copper, lead and zinc, are rendered more available and more toxic by declining pH, but pH is but one of a complex set of toxicity-controlling factors.

In all these cases the effect of hydrogen ions is probably directly on the mineral lattice concerned, emphasising the indirect relation to plant physiology.

Figure 11.2: Aluminium availability in tropical soils with pH. Data from Sanchez (1976)

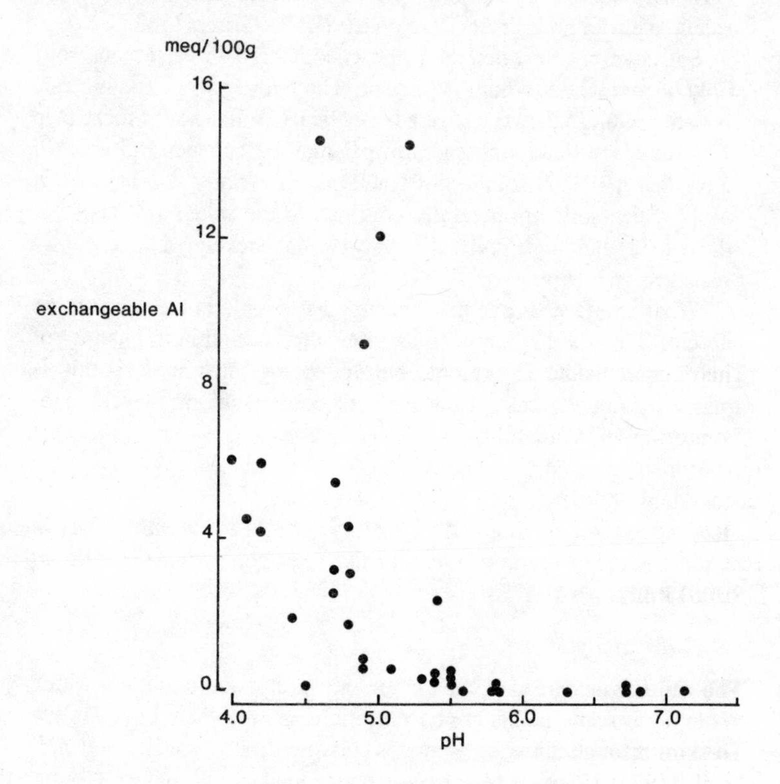

12

Experimental approaches to the study of soil variables

This chapter is divided for convenience into the following topic areas: substrate analysis, plant analysis, field environmental monitoring; perturbation experiments; greenhouse and other culture experiments; ecophysiological and biochemical studies; and the ecosystem approach.

SUBSTRATE ANALYSIS

The objectives of analysis

To skim through a technical volume on soil analysis such as Allan *et al.* (1974) *Chemical analysis of ecological materials*, or Hesse (1971), *A textbook of soil chemical analysis*, can be a confusing experience without proper orientation (see Further Reading). Equally confusing is the reading of research papers on soil–plant relationships or biological floras in which soil analysis data are recorded. The motivation for analysis is frequently taken for granted or otherwise remains unclear. Agriculture has laid down analytical guidelines to determine the suitability of a soil to a particular crop; and the fertiliser requirement for optimum yield of that crop. In contrast, there are different approaches to soil description for the purposes of ecological interpretation:

(a) Pedological or geomorphological description of soil types within a landscape in order to establish broad correlation with vegetation type. These descriptions may form part of the introduction to an ecological account. They will obviously entail description of more than the rooting zone.

(b) Detailed chemical description of a variety of chemical and physical characteristics, without being assertive about causal relationships, e.g. biological floras. Soil data bases of this type are valuable materials towards building hypotheses on ecological cause and effect.

(c) Soil studies as part of an ecosystem analysis, i.e. to establish the magnitude of the soil pool for elements of interest.

(d) Chemical analyses which aim to isolate, define and quantify critical characteristics of the soil environment. These may be with reference to control of species distribution or growth performance, e.g. nutrient availability or concentration of toxic ions. In many hypothesis-testing studies, it can be important to eliminate variables and narrow the field for intensive investigation.

It is not necessary to compare the merit of each approach, but it is vital to examine intellectual motives before embarking on soil analysis. Without critical examination, results obtained may be useless, and costly waste of time and money will result.

Motives for soil analysis in ecology

(1) To give a preliminary description of the soil as an environment for root growth and metabolism, seed germination and the various activities of other soil organisms.

(2) To formulate hypotheses on environmental interactions with biota which may limit biological activity.

(3) To test hypotheses by various means including detailed observations, enrichment or tracer studies, exclusion of grazers, etc.

(4) To measure chemical and biological activity in soil quantitatively with a view to interfacing with other data, as in ecosystem studies.

(5) To permit comparison with other sites as relevant. This may widen an examination to include investigation and interpretation of soil development.

Sampling, storage, etc.

The processes of sample collection and storage will have a very great influence on the results obtained from soil analysis. This may well be the most critical step in the chain from the field to a sheet of data for interpretation. In terms of financial cost, physical effort and logistic difficulty the sampling of the field situation is also the most taxing part of this procedure. Thinking about sampling should take into account the following:

(a) Scientific reasons

Decide how your situation fits into the scheme of motives. Is the occasion a preliminary reconnaissance, the start of a regular series of observations, or the establishment of an experiment (to name just three possibilities)? A checklist of what is to be measured or determined analytically should be decided before embarking on a collecting expedition. The sampling pattern should be decided as clearly as possible. A choice may need to be made between transect or grid methods for investigating heterogeneity; preliminary random sampling of potential study sites; and stratified random sampling where study units are known to possess different degrees of variability.

(b) Logistics

The type of analysis to be carried out will determine what needs to be sampled. For a determination of particle size structure, supporting soil characteristics and nitrogen mineralisation rate of the rooting zone of a selected site, three similar samples would be collected. One would be dried and reserved for particle size analysis. The second would be subsampled fresh for pH determination, then dried and 2-mm-sieved for supporting soil characteristics. The third sample would be carefully subdivided for the nitrogen mineralisation determination, ensuring that appropriate storage temperatures and humidity were observed. This example makes it clear that a logistic penalty must be paid for detailed examination.

The nature of the problem at the site level will be the other determinant of sample numbers. This means consideration of sites, sampling patterns and replication for statistical purposes. Knowing the order of magnitude of sample numbers is not only of importance in calculating how many containers to take or what weight must be borne from the site. It will also determine the time and cost of subsequent laboratory operations. If a large study is contemplated, a box diagram of the procedure indicating the treatment of samples and the time required for each stage may help planning.

(c) Mechanics of sampling

Given the layered structure of soil, it is important that layers are not over- or underrepresented in a sample. Samples collected with spades or trowels tend to overrepresent upper layers. A very convenient and effective soil sampler is a stainless steel corer 2.5–3.0 cm in diameter. A 1-cm slot cut into one side will reduce compression, and a removable tommy-bar placed through holes in the top facilitates penetration of all but the most stony soils. Such a corer will ensure collection of a volumetric sample in most cases. It fails with compressible, fibrous organic soils, when a sharp knife may be used to cut a defined cubic sample. Most methods fail with very stony soils, and it may take pains to sample fine earth uniformly from the rooting zone. Fresh soil samples are conveniently transported from the field in plastic film or stout paper bags.

PLANT ANALYSIS

Compared with substrate analysis there are, as yet, fewer guidelines concerned with the role of plant tissue analysis in ecology. In agriculture the routine application of tissue analysis is confined to very few widely grown crops.

The most important application of tissue analysis in ecology is in ecosystem studies. In such studies, serial harvests of plant material are stratified to sample plant parts. Budgets for economy of nutrient elements can thus be compiled for species on a seasonal basis. This is probably the best method of intercomparison available at the present time for ecological purposes. For reasons explained in Chapter 2, it is not valid to intercompare plant tissue analyses except under closely defined circumstances. An exception to this is the recognition of ion accumulation where toxic elements are concerned. Analysis of leaf tissue may be a useful way of screening for absorption of elements such as lead, copper, zinc, cadmium, fluorine and selenium.

The most difficult technical problem to overcome is recognition and control of surface contamination of tissues by dust or soil particles. This may be accomplished by making careful observations of the quantity of ash of a tissue, even inspecting with a microscope for traces of soil or dust. Carefully controlled sampling and analysis of specimens with minimal contamination will provide standards.

In some cases high values for predominantly soil elements, e.g. iron or titanium, may provide markers of soil contamination. These are not especially useful if the analyte and potential surface contaminant are the same, e.g. lead in dust on the surface of a plant in a lead assimilation experiment.

FIELD MONITORING OF THE SOIL ENVIRONMENT

Many features of the soil environment are known to fluctuate under field conditions. To study them field observations must be structured in a time series. The periodicity of measurements will vary substantially with situation. A simple example is the diurnal fluctuation of soil temperature which must also be viewed in a seasonal context. In the case of soil moisture there are also diurnal and seasonal rhythms, but it may be necessary to seek critical values in the range, i.e. for how long soil ψ_m is below wilting point for a particular species.

The salinity of intertidal saltmarsh soils may be even more complex. A 'normal' range is determined by the diurnal tidal cycle of spring tides. Extreme salinities are generated either by dilution as a result of rainfall or by high evapotranspiration.

To determine supply of nitrogen, monthly measurements of nitrification would show predictable fluctuations which should be evaluated with respect to growth of vegetation. The pattern of soil aeration will include seasonal variation combined with changes in intensity and duration distributed through the soil profile and aggregate structure.

Measurement of fluctuating soil environmental conditions may take at least three different forms (Table 12.1)

(a) Measurements requiring serial collection of samples for laboratory examination. This approach demands sampling with minimum disturbance of the variables of interest and rapid analysis. Good examples are most ion availability measurements and detailed studies of solute potential.
(b) Characteristics monitored *in situ* by manually operated field instruments. This requires a robust but sufficiently sensitive sensor which may be either inserted temporarily into the soil, e.g. for determination of redox or oxygen diffusion rate, or permanently emplaced. Insertion or emplacement of sensors must not cause undue alteration of soil conditions. It may be a

Table 12.1: Field monitoring of soil environment

	Sample to laboratory	Manual field instrument	Automated field method available
Soil temperature	Not applicable	+	+
Soil matrix potential	+	+	–
Soil aeration:			
(a) water-filled pores	+		
(b) O_2 diffusion rate		–	–
(c) CO_2 in soil		–	
(d) redox potential		+	
Soil solute potential	+	+	
Soil conductivity		+	–
Available nutrients:			
(a) nitrification	+		
(b) ammonification	+		
(c) phosphate	+		
(d) cations	+		
(e) hydrogen ion	+	+	

(+, fully developed routine procedure; –, prototypes described)

practical proposition to emplace either water potential or temperature sensors in soil and make records with a portable meter at convenient intervals.

(c) Measurements made by *in situ* sensors with continuous automated recording. In many cases this may seem to be an irrelevance for technical, cost and security reasons. However, this is a rapidly developing field and many technical and cost barriers to the use of such techniques may disappear. Sensors are available for temperature, water potential and soil aeration characteristics. Data logging techniques have been developing steadily, and the most recent developments are in the microcomputer processing and analysis of data from automatic systems. The most needed development is the provision of nutrient ion sensors for field emplacement. For example, to have a continuous record of nitrate and ammonium concentration in soil liquid phase would have very great ecological (and agricultural) value.

PERTURBATION EXPERIMENTS

Some ecological hypotheses can be tested by direct experimentation. Experimental manipulation of the field environment is the most obvious mode of experimentation, but its applications are limited. The main applications are:

(a) investigating hypotheses of nutrient limitation by experimental fertiliser addition;
(b) investigating the effects of predation by herbivores, usually by exclusion experiments;
(c) miscellaneous experiments including artificial watering, burning, forest cutting or drainage.

The first application of any experimental treatment may well be empirical, in the sense that theoretical guidelines for treatment may be absent. For example, Jeffrey and Pigott (1973), when applying nutrient addition treatments to grassland in upper Teesdale, used one dose rate, 5 g P m^{-2} and 10 g N m^{-2}. In the case of the phosphate-P dose rate, this amounted to about 100 mg P per litre of soil or approximately doubling the total phosphate content. However, for short-term phosphate availability, because $CaHPO_3$ was added to the calcareous soil, an uncontrolled pulse of available phosphate was probably generated. One suspects that in many ecological perturbation experiments the exact effect of the treatment is seldom understood. In an ideal nutrient addition experiment, the fate of all added elements should be accounted for. This is particularly necessary when non-response is elicited. Addition of soluble nitrogen to saltmarsh in Norfolk (UK) (Jefferies 1977; and Dublin M.J. Sheehy-Skeffington and D.W. Jeffrey, unpublished data) gives rise to no measurable growth response. This is not readily explained as the vegetation has other signs of nitrogen limitation. Possible reasons include growth limitation by salinity; removal of soluble nitrogen by leaching by tide water; and denitrification. This example indicates that nitrogen addition experiments may give rise to interesting difficulties. The apparent simplicity of the classic experiment of Willis (1963), who demonstrated that nitrogen limited plant productivity in coastal dune grassland, may be a rare phenomenon.

GREENHOUSE AND OTHER CULTURE EXPERIMENTS

It is tempting to abandon the field in hypothesis testing and adopt more simple and more controlled units of study. The main advantage is the capacity to isolate particular variables at will and to manipulate them in the absence of uncontrolled variation. A number of cautionary observations may be made as guidelines towards the confident use of this approach:

(a) Ensure that the range of amplitude of experimental treatments is realistic in terms of the field. For example, in dealing with saltmarsh soil salinity, experimental treatments should relate to the −2.4 MPa osmotic potential of seawater. However, it must be noted that the uppermost zones of saltmarsh soil salinity may seldom reach this value but in some parts of the marsh system it could be as low as −4.8 MPa. Field assessment of amplitude should precede experimental manipulation where possible.

(b) Second-order effects, eliminated in experimental systems, may be important. A common feature of water-culture grown plants is an absence of mycorrhizas, but massive bacterial films may build up on root surfaces.

(c) Experiments are usually carried out on young seedlings:
   - (i) Selection of ecotypes may eliminate unsuitable genotypes in the field. There is a case for use of clonal material in experiments.
   - (ii) Conditions for seedling establishment may differ from long-term survival; in particular, nutrient carry-over from the seed may have long-lived effects in terms of the length of the experiment.

(d) Changes may occur in soils removed from the field. Soil structure will obviously be destroyed if soil is removed, transported, homogenised and placed into containers. Soil fauna may be moved or their activity modified.

(e) Greenhouses in particular may generate critically different temperature regimes, with high daytime air and substrate temperatures. Under more closely controlled growth cabinet regimes, light levels may be unsatisfactorily low, or of unsuitable spectral balance.

(f) Where possible measure the manipulated variable as closely as possible. For example, it is quite difficult to adjust pH of a medium accurately in the presence of a living plant. Therefore it is pragmatic to adjust within possible limits, but then to monitor the experimental system closely.

On the other hand, the infinite possibilities for varying soil conditions in relation to one or more species make small-scale experimentation virtually indispensable to plant ecology. At our disposal we have the possibility of using field-collected soil, synthetic growth media or water culture. The use of greenhouses or growth chambers greatly extends the time available for experimentation and permits the elimination of climatic constraints on growth.

ECOPHYSIOLOGICAL AND BIOCHEMICAL STUDIES

At the heart of many ecological phenomena lies a particular physiological or biochemical mechanism. Examples already encountered include C4 and CAM photosynthesis; synthesis of secondary metabolites protective against herbivores; and phytase activity of mycorrhizal roots. It is obviously necessary to experiment at this level of study from time to time. Because of the very great potential diversity it is difficult to make up useful ground rules. Nevertheless, it is generally important to realise that, when the object of study contracts from a complex field situation to a single tissue or tissue extract, the original broad problem should not be forgotten. The physiological or biochemical answer to a problem should really represent yet another ecological hypothesis. This means that the laboratory results must be tested in the field. For example, there are many potentially toxic or repellent compounds to be found in plants. Laboratory studies may also indicate that these have a direct effect on potential predators. However, without direct observation of the effectiveness of protection, the idea remains hypothetical.

Without this last rigorous phase, we run the risk of constructing an 'ecology of wishful thinking' composed of ingenious but untested mechanisms. Among the examples discussed in Part III are some stimulating but as yet untested hypotheses of this type.

ADAPTING THE ECOSYSTEM APPROACH FOR VEGETATION STUDY

The fundamental questions asked by ecosystem studies concern the utilisation of environmental resources by living organisms. Although these may ultimately entail the integration of food-web studies plus consideration of transfers into and out of the system, vegetation-oriented work can commence at a simple level. For non-woody vegetation this may be a relatively straightforward procedure. In my view it is worth testing the hypothesis that the quantity of major nutrient elements contained in the vegetation of ecosystems and recirculated on an annual basis is of great importance in understanding the evolution of ecosystems. For woody ecosystems, the procedure entails a more elaborate system of subsampling and allometric approximation (Newbould 1967). The value of such studies is demonstrated in Chapter 15.

Determining the deployment of nutrient elements in non-woody vegetation

This determination and its interpretation may be seen as a set of successive approximations. The first of these is the assembly of data representing the maximum live standing crop. In the north temperate zone this usually develops in late July or early August. The procedure leading to nutrient deployment as grams of nutrient element per square metre is set out in Table 12.2. Apart from the obvious sampling and data-handling questions, the most difficult technical problem is dealing with root biomass. The next step is to determine the deployment between vegetational components, living and dead plant material and species. It must be realised that resolving the nutrient deployment picture to this level of detail for one harvest may not be especially illuminating in light of (a) the pattern of seasonal senescence; (b) the relative contribution of particular species; and (c) the elemental composition of species. These are certainly altering continuously throughout the annual cycle, and the fullest picture is obtained by the next step. This is to repeat the exercise at appropriate intervals through the season. For a single worker this is a relatively ambitious undertaking but very rewarding in terms of the depth of understanding obtained.

Table 12.2: Stages in determining the deployment of nutrient elements in non-woody vegetation

First approximation

1. Select appropriate quadrat size and determine degree of replication needed. Decide nutrients.
   Options:
2a. Harvest above-ground standing crop by clipping to soil level.
2b. Obtain root sample by corer or other suitable technique. Wash soil from root samples.
3. Dry harvested material.
4. Weigh dried material. Calculate biomass, kg m^{-2}.
5. Homogenise by milling or other means before or after subsampling.
6. Prepare samples for analysis by wet ashing.
7. Determine elemental concentration by appropriate analytical procedure. Calculate concentration (mg element kg^{-1} d.w.). Multiply by biomass to calculate deployment of nutrient element value (mg m^{-2}).

Second approximation: categorisation

1. Determine sampling procedure, degree of replication, elements to be determined, and species or categories for sorting.

   Options:
2a. Sample roots and tops together. Dissect and wash roots while attached. Sever before drying.
2b. Cut above-ground parts and sort. Sample roots and sort using labelled specimens to identify.
2c. Cut above-ground parts and sort. Sorting of roots to species level may not be possible. Attempt to determine living vs. dead roots.
3–7. As first approximation.

Third approximation: seasonal dynamics

In addition to the above, decide sampling interval. From the biomass maxima and minima, and tissue concentrations of elements, turnover estimates may be calculated. The most simple of these is: (maximum biomass × tissue concentration) − (minimum biomass × tissue concentration) = turnover. This is expressed as weight of element per unit area per year. More realistic estimates may be determined by aggregating inputs and outputs over a series of shorter time periods. Outputs will include losses due to senescence, decomposition and grazing. See Milner and Hughes (1968) for further details.

Part Three

Interactions in the real world. Some case histories

The case histories are selected primarily on the basis of their interest to the author. This gives a certain licence to comment and speculate on other authors' work. Scientific method provides a methodology for testing hypotheses, but the selection of hypotheses is to some extent the product of intuitive and imaginative speculation.

The cases are arranged in a rough order of complexity, from a pair of autecological investigations to the substantial complexities of tundra, saltmarsh, limestone and serpentine vegetation.

In each case use will be made of ideas introduced elsewhere in the text. One applied topic, restoration of derelict land, is introduced overtly, but in all other cases applications are evident. Ecological management depends on cases such as these.

It is to the author's regret that a truly tropical example is not included, but the thinking behind most of the cases is valid more widely than in the region of its origin.

13

The autecology of two contrasting species

These two case histories may be read as an apologia for the sometimes self-indulgent practising of autecological investigation. Studies of single species, unless carried out in a rigorously comparative frame of reference, may not repay the effort expended. In these two cases the species selected are representative of wider themes, and each study ought to stimulate further work.

A coincidence, but a reason for selecting these two cases to begin with, is that a powerful single factor appears to predominate in both, namely soil phosphate availability. This should be regarded as a rare event, as the ultimate causes of ecological phenomenon are almost invariably complex.

THE ECOLOGY OF THE STINGING NETTLE, *URTICA DIOICA*

The stinging nettle is widely distributed in temperate regions and must rank as one of the most popularly recognised herbaceous plants. Before the widespread use of flax and cotton it was used as a source of fibre, and its young shoots are still eaten as nutritious spring greens.

Urtica dioica was also one of the first single species subject to ecological study. In 1883 Molisch demonstrated that it was one of a group of species which contained free nitrate in its tissues. In 1921 Olsen published an account of the ecology of nettle in Danish woodlands. Nettle soils were seen to be rich in most nutrients, but special attention was devoted to demonstrating the apparent importance of nitrogen. Not only were nettle tissues rich in nitrate when compared with *Deschampsia caespitosa*, but also nettle soils had comparatively high nitrification rates. Growth in water cultures showed also that nettle was capable of response to a wide range of nitrogen concentrations. In the light of this evidence it seemed reasonable to apply the term 'nitrophilous' (nitrogen loving) to the plant. It was widely assumed that both its natural distribution in woodland glades and its distribution in association with farmyards, byres, latrines and rubbish heaps was conditioned by high nitrogen supply.

In an investigation of the comparative ecology of woodland herbs in a limestone dale in Derbyshire (UK), Pigott and Taylor (1964) encountered nettle separated ecologically from *Deschampsia caespitosa* and another woodland herb, *Mercurialis perennis*. The nettle plants clearly had higher tissue concentrations of nitrogen and phosphorus than the other two species. When the nitrogen status of soils associated with all three species was compared, the differences were slight. Interpretation of results of nitrification experiments on sieved soil was not straightforward, however, because of variable stone content. Correcting for stones and expressing results on a square-metre basis did indicate a slightly better nitrogen supply to nettle.

Pigott and Taylor's next investigation was to explore the growth of nettle seedlings on a series of six soils from Derbyshire and Cambridgeshire (UK) sites associated with *Deschampsia caespitosa, Mercurialis perennis* or *Urtica dioica* as natural dominants. In these greenhouse experiments, nitrogen or phosphate was added to small pots of each soil 5 cm in diameter and containing 500 cm^3.

The results of two of these experiments are presented in Figure 13.1. The overwhelming response of nettle seedlings to phosphate addition on soils where nettle is not found, and the virtual absence of any response to added nitrogen, is a strong indication that phosphorus supply is a limitation to colonisation of non-nettle soils. In further experiments with soils where nettle grows strongly, little or no response to phosphate was recorded. On soils where mature plants of nettle are weakly grown, seedling growth was checked unless phosphate was added.

The cycle of experiments was completed by a pair of field experiments on *Mercurialis* sites in which 5 g phosphate per m^2 was applied and nettle seed sown. Here the response to phosphate occurred strongly where light was adequate, and nettle plants overtopped *Mercurialis*.

The ecological behaviour of *Urtica dioica* is thus largely explicable. Essential soil conditions are for a large supply of available phosphate and reasonably high availabilities of other major

Figure 13.1: Results of two greenhouse experiments in which nettle (*Urtica dioica*) seedlings were grown on 'non-nettle' soils with added nitrogen or phosphate

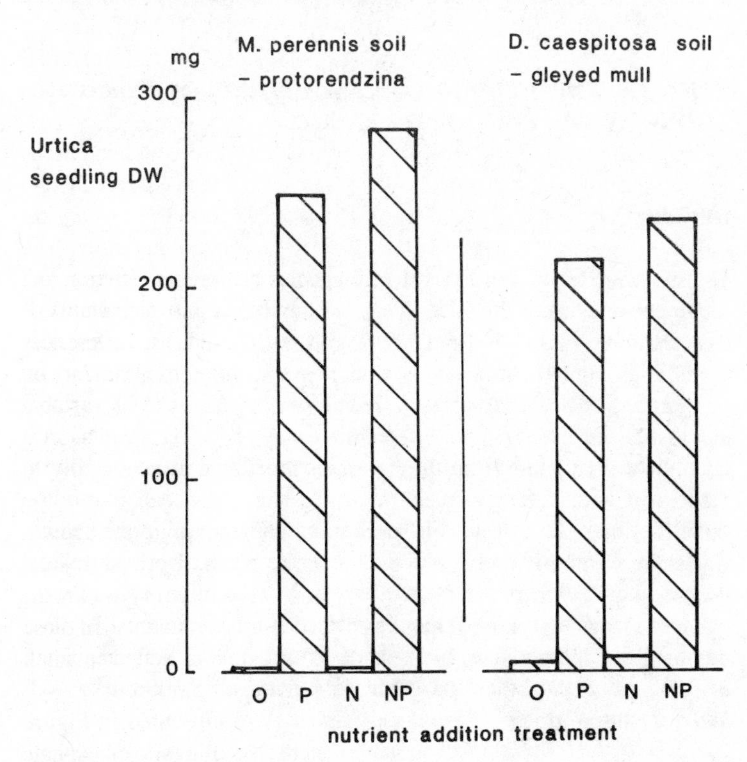

nutrients. It is a plant often associated with burning of vegetation and lairage sites for sheep or goats. In the latter context it is commonly encountered unexpectedly in montane areas at altitudes up to 3000 m.

The final questions about *Urtica* must concern the metabolic details that are associated with the demand. Storage of phosphate undoubtedly occurs (Nassery and Harley 1969) and the rhizome system probably has a reserve of phytin, which the fast-growing shoots utilise in spring when growth commences. A last question, unanswered as far as I know, is whether other members of the genus behave in the same fashion. *Urtica urens* is certainly a plant of bird islands and agricultural waste places, but is an annual. Companion species in Europe include the woody plant *Sambucus niger* (elder) and the herb *Galium aparine* (goosegrass). These would seem to be candidates for similar patterns of investigation.

'High phosphate demand', as illustrated by nettle, is a curious phenomenon when viewed against the economies referred to already, and to be illustrated further.

KOBRESIA SIMPLICIUSCULA — CASE HISTORY OF AN ARCTIC ALPINE AT THE EDGE OF ITS RANGE

Introduction

Unusual vegetation and several rare species are special features of the upper reaches of the River Tees, which drains part of the north-east Pennine range in England (Pigott 1956). The rare species include a number of arctic–alpine species, in particular *Dryas octopetala, Gentiana verna, Tofieldia pusilla* and *Kobresia simpliciuscula*. Only one, *Kobresia*, may be regarded as an ecological dominant. Here these species are found at about 500 m associated with outcrops of thermally metamorphosed Carboniferous limestone. As a result of heating and hydrothermal action, the limestone weathers to give individual calcite crystals reminiscent of coarse sugar. Barium sulphate (barytes), lead sulphide (galena), calcium fluoride (fluorspar) and hydrated iron oxide (goethite) are companion minerals, usually as enriched veins. Acid soils on glacial drift surround the limestone outcrops, giving rise to complex soil and vegetation gradients.

Hypothesis development

The key question to be addressed was how a small group of species, in particular a slow-growing sedge, managed to survive as relicts for some 10 000 years in upper Teesdale. The principal factor threatening survival was hypothetically the competition from upland grassland species which surround the site. Possible reasons for the failure of common grassland species to cover this patchily vegetated area are:

(a) Climatic effects. In spite of the relatively low altitude, the climate of the site has been described as having much in common with southern Iceland with a long period of winter snow cover and a cool and cloudy summer.

(b) Grazing, including about 700 years of sheep grazing and intermittent rabbit grazing.
(c) Soil fertility, bearing in mind especially the experiments of Willis (1963) who demonstrated the responses to nitrogen of coastal dune vegetation.
(d) Unknown effects of the metamorphosis of Carboniferous limestone and its mineralisation.

The investigation (Jeffrey 1971; Jeffrey and Pigott 1973)

It was possible to undertake perturbation experiments on this site because of the construction of a water storage reservoir. In 1971 the lower outcrops investigated in this study were flooded. The main constraints on the investigation were a two-year time span and the relatively small areas of *Kobresia*-dominated vegetation available for investigation. The final scheme for the investigation entailed:

(a) Fencing the site against sheep.
(b) Design of a simple fertiliser application experiment applying nitrogen as ammonium nitrate and phosphate as $CaHPO_4$. Since destructive sampling was not feasible, monitoring of responses was achieved by using a point quadrat system. This method satisfactorily estimates the cover of species in short grassland, especially if comparisons are made at similar times of year.
(c) Investigating soil chemistry.
(d) A miscellaneous set of greenhouse and laboratory experiments.

Results

The striking and surprising result of the fertiliser application was the massive response to applied phosphate. Expressed simply in Figure 13.2 it can be seen that grasses, especially *Festuca rubra*, increase their cover, and a further response occurred to the nitrogen-plus-phosphate interaction. Soil phosphate availability appears to be a major factor restricting competitiveness of grasses. When all the available point quadrat data are summarised it may be seen that cover of *Kobresia* decreases as total cover increases. We did not, however, observe elimination of *Kobresia*.

The soil investigation took the form of investigating the chemistry

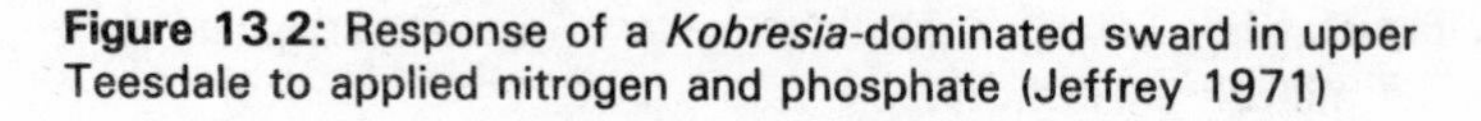

Figure 13.2: Response of a *Kobresia*-dominated sward in upper Teesdale to applied nitrogen and phosphate (Jeffrey 1971)

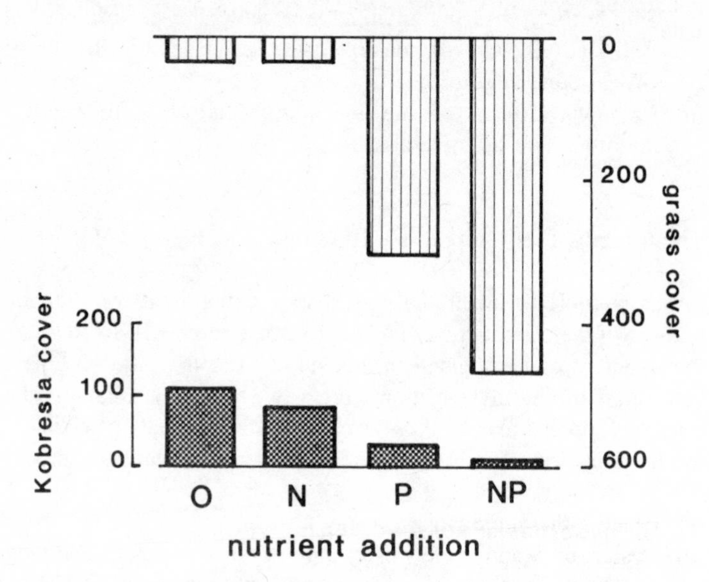

of soil samples from experimental sites and other *Kobresia* localities. Samples were also collected from neighbouring communities, sometimes structuring sample collection in short transects across the sharp vegetation boundaries on the site.

Data on major nutrients and trace metals did not immediately fall into an intelligible pattern, even when expressed on a soil volume basis. However, when exploring correlations between soil characteristics, a remarkable positive correlation was noted between inorganic phosphate and total lead in the *Kobresia* soils (Figure 13.3). This suggested that lead in soil was acting as a depressant of phosphate availability. Quantifying this, it seemed possible that all the phosphate–lead interactions were in effect producing soil conditions approximating to 50 mg/phosphate-P per litre of soil. This concentration is low by any standards, irrespective of availability considerations. In order to test the plausibility of this hypothesis, a direct physical simulation of the soil system and its effects on the growth of grasses was attempted. A series of four sand cultures were set up in which two different concentrations of solid-phase phosphate (hydroxyapatite) were interacted with solid-phase lead as finely ground galena (PbS). Calcium carbonate was added to

Figure 13.3: Correlation between inorganic phosphate in *Kobresia* soils and total lead (A). Growth of *Festuca ovina* on substrates containing phosphate as hydroxyapatite and lead as galena (C) (Jeffrey 1971)

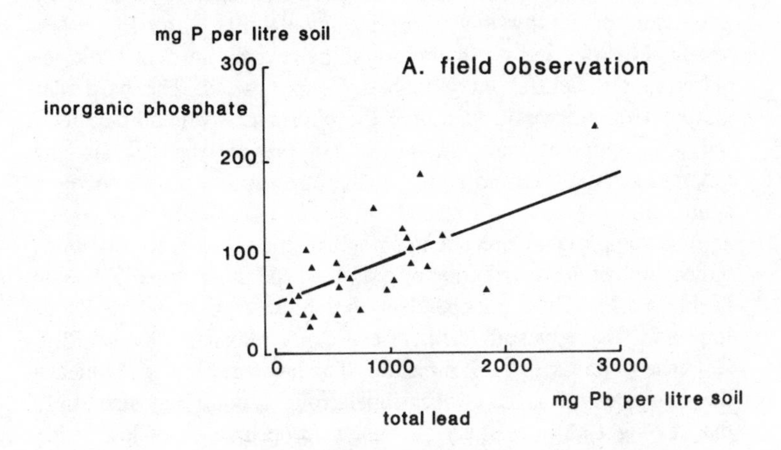

B. hypothesis: soil lead depresses phosphate availability

C. experimental verification

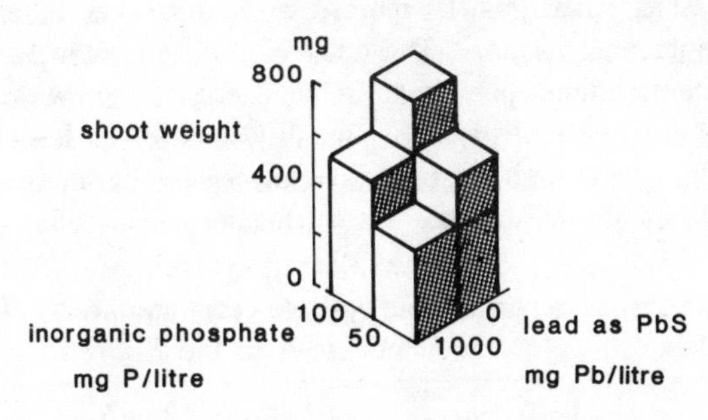

maintain a high pH in simulation of the calcareous field substrate. All other nutrients were added as a water culture. The results support the hypothesis of a phosphate–lead interaction. They also support the validity of regarding the soil matrix as a three-dimensional grid with the spacing of slowly diffusible elements an important characteristic in the control of their availability.

Conclusion

The niche that *Kobresia simpliciuscula* occupies in Teesdale is flushed with calcareous water and protected against invasion by grassland species by low phosphate availability. Growth of this species is slow, and not markedly stimulated by nutrient additions either in the field or in greenhouse experiments. The particular climatic regime would reinforce the phosphate availability effect, but it is apparent that climate did not prevent the response by grasses, such as *Festuca rubra*, which are successful in a lowland situation.

It is tempting to erect a hypothetical model of soil conditions which control the distribution of grassland species in upper Teesdale (Table 13.1). Three independent variables seem to stand out as important, namely soil fertility, especially phosphate availability, soil base richness and soil moisture. The last factor certainly means the presence or absence of occasional drought, but it may also imply that 'drought absence' may be related to occurrence of low redox episodes.

The relationships between the niche types are conveniently expressed in Figure 13.4, using a graphic form employed by van Leeuwen (in van de Maarel 1971). In constructing this matrix it is emphasised that this is merely an aid to visualising possible environmental variants. The axes symbolise continuous variables, and intermediate positions, rather than sharp boundaries, are probably more realistic. The 'fertility' axis is the least understood and most oversimplified of the characteristics. It needs extending to include meadows and pastures on former forest soils.

Table 13.1: Soil factors and grassland dominants in upper Teesdale

Fertility	Base richness	Soil moisture	Dominant
Low	High	High	*Kobresia*[a]
High	High	High	*Festuca rubra/ Agrostis stolonifera*[a]
Low	Low	High	*Nardus stricta*[a]
High	Low	High	*Deschampsia caespitosa*
Low	High	Low	*Sesleria*[a]
High	High	Low	*Cynosurus/Lolium*
Low	Low	Low	*Agrostis tenuis*[a]
High	Low	Low	Not represented?

[a] Encountered directly in investigations on the Teesdale site.

Figure 13.4: Diagram illustrating the hypothetical relationships between grassland types in Teesdale in terms of soil fertility, base richness and soil moisture. It is emphasised that these are comparative rather than absolute characteristics. The diagram is based on the independence of these three characteristics

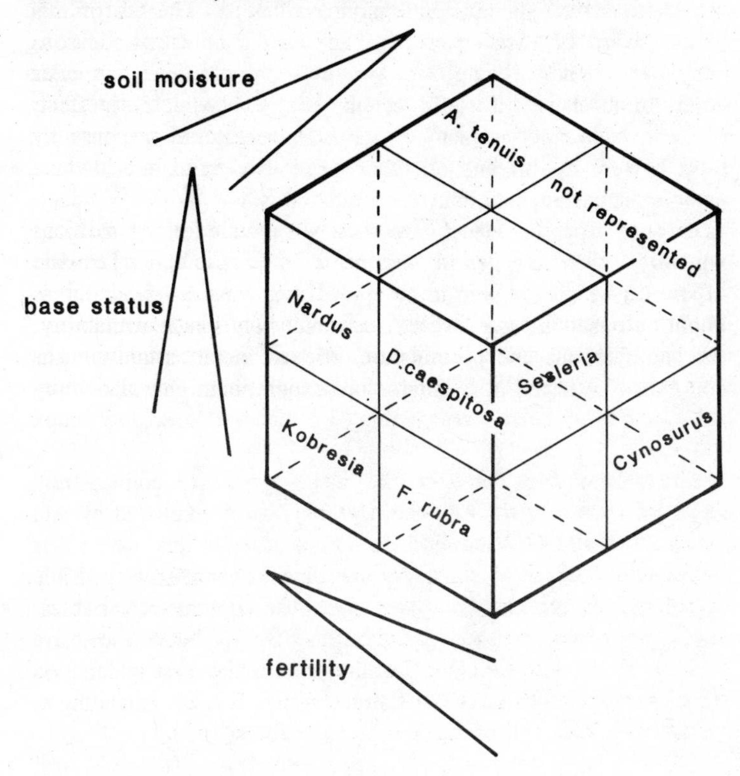

Separation of soil characteristics regulating plant performance in an upland grassland

Another approach to separating the factors responsible for performance of grassland plants has been published by Rogers and King (1972). Their unit of study was an upland grazing in Scotland, and their aim was to determine soil factors giving rise to distribution and performance of particular species. This knowledge could then be utilised in pasture management. Detailed environmental measurement accompanied quantitative measurements on sward composition. The soil data were then inspected statistically for correlations between environmental factors. Two groups of independent factors emerged, those connected with base status and those describing soil aeration.

The sward composition data were then presented, species by species, as an 'isonomic' diagram. Here two axes are erected, representing base status and aeration coordinates. The contribution to the sward of each species is then plotted as a contour map, computer methods being used to draw the isonomes of equal contribution. This is a useful technique for exploring the influence of a pair of interacting factors on a single species. However, users must beware of this and all other correlative techniques. As the authors themselves point out, a correlation is not complete evidence as to cause. Correlative techniques are best used to refine hypotheses which may then be tested in other ways. For example, performance of species which are seen to be optimal under base-rich conditions might be observed in a 'before and after' field liming experiment. All combinations and permutations of base richness and aeration could also be tested in a suitable pot experiment.

14

Restoration of derelict land

The insertion of an applied topic is calculated to emphasise that the same processes of analysis and experimentation hold good for both pure and applied ecological studies. Almost all fields of ecology have an applied aspect, something that needs emphasis in today's world. This particular topic also serves to introduce new subject matter and is in itself a stimulating field (Bradshaw and Chadwick 1980).

Some human activities damage vegetation and soil so severely that the land is rendered useless for an intolerably long period. These activities include mining and quarrying; clearance and regrading of large construction sites; dumping of waste materials; and demolition of obsolete industrial installations. Land dereliction is usually accompanied by serious environmental, economic or social problems including soil and water pollution, erosion and under-utilisation of resources. This is formally recognised in many parts of the world by legislation requiring the restoration of derelict land. The task of restoration requires application of a wide range of knowledge of soil–plant relations. The single principle to be adopted is that land restoration entails reconstruction of the entire ecosystem. We may concentrate on soil, plants and microorganisms initially, but other aspects of the food web need to be kept under review.

A BROAD ANALYSIS

An essential practical requirement for any application is a structured approach as in Figure 14.1. This scheme was originally designed as a means for communicating with engineers and administrators, but it will serve as a basis for discussing some examples.

Figure 14.1: Scheme for the planning of rehabilitation for a derelict site

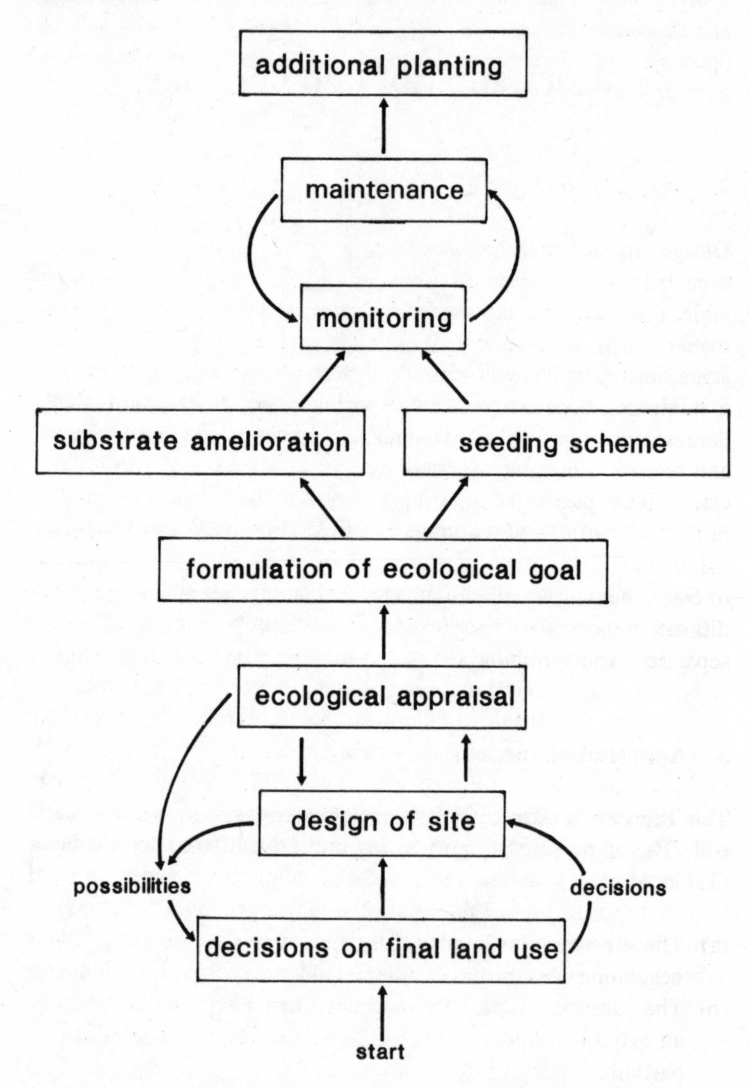

1. Decision on land use

The type of decision to be made will vary enormously between situations. Under some legislative codes dealing with extractive industry, restoration of 'native vegetation' is mandatory. In contrast a wide range of options may be possible in an urban setting, e.g. sports grounds, amenity woodlands or industrial uses. It is possible, and desirable, in some cases to re-create agricultural land — after opencast coal mining (strip mining), for example. Afforestation is another important land-use option for many situations.

2. Design of operation

Design means that the engineering possibilities are explored, especially with respect to topography, e.g. slope, aspect, water table, drainage. If a mining waste situation is being examined, and matters such as slope and orientation of dumps are at a planning stage, unfavourable north- or south-facing slopes can be reduced to a minimum. Decisions can also be made to include water-filled depressions in a design scheme or fill them. A major consideration, and often a constraint, will be cost. This may have to be met as a charge on the operation causing dereliction. In mining this may be in the order of 1% of the value of the product. Alternatively, costs may be set against the future value of the restored site. The question of cost usually bears directly on the soil–plant manipulations that are ultimately possible. Large-scale earth-moving operations, when separated from productive activity, are generally uneconomic.

3. Appraisal of site and substrate

This exercise is identical to the ecological appraisal of a vegetated soil. The appraisal may lead to a number of different conclusions (Table 14.1):

(a) The substrate closely resembles a non-extreme natural soil, and vegetation may be easily established.
(b) The substrate, especially in the context of the site, lies within an extreme range, e.g. high or low base status, overdrained or partially waterlogged.
(c) Extreme problems exist. These may include hyperacidity, metal toxicity or salinity.

Table 14.1: Appraisal of main substrate characteristics encountered in land restoration

Problems	Solutions
(1) High compaction (common)	Deep ripping or cultivation
(2) Low phosphorus or potassium (occasional)	Add fertiliser or nutrient-rich waste. Inoculate with mycorrizal symbionts
(3) Low cation exchange capacity (common)	Add organic matter — sewage sludge, peat, agricultural or food-processing wastes
(4) Low nitrogen (inevitable)	(a) Give a small priming dose of nitrogen (b) Sow legumes, ensuring they are inoculated (c) Ensure free-living nitrogen fixers are present by topsoil inoculation
(5) Waterlogging	Determine cause and establish: (a) Will drainage or cultivation lower water table? (b) Is it permanent or periodic? (c) Should a water feature be included?
(6) Slope instability	Reprofile to decrease angle of slope below angle of repose
(7) Large particle size	Examine pocket planting options
(8) Presence of pyrite leads to hyperacidity (occasional)	Very complex problem — see below
(9) Presence of high concentrations of toxic metal ions	Another complex problem — see below

4. Formulation of an ecological goal

The appraisal in Table 14.1 will lead naturally to the development of ecologically compatible goals. For example, limestone grassland was chosen as a suitable model for calcareous mine wastes in western Ireland, although these wastes were seriously metal contaminated (Jeffrey, Maybury and Levinge 1975). Non-toxic overburden from opencast coal mining could be readily reconfigured to permit re-establishment of grazing land. More acidic, less intrinsically fertile coal wastes have also been afforested with a minimum of reconfiguration, e.g. Bradshaw and Chadwick (1980).

5. Development of amelioration programme and seed mixture

It is implicit in the word 'development' that site treatment and planting schemes need to be explored experimentally rather than followed from a suitable manual. These are a pair of exercises, usefully carried out in parallel. From the ecological appraisal, a range of species is suggested. From the site and substrate analysis, possible treatments are suggested. Costs and opportunities are also variables to be handled by the experimenter. Opportunities include the availability of low-toxicity substrates such as overburden or newly stripped soil, the use of temporarily underutilised heavy machinery or the seasonal availability of labour.

Field pilot experiments are generally the most common and cost-effective approach, combining both soil treatment and planting scheme. Typically a series of soil treatments, such as alternative surfacing substrates, ameliorant or fertiliser additions, cultivation or irrigation techniques, are prepared on a small scale. These may then be tested with a hand-sown seed mixture and effects on establishment will be determined in a few weeks. Within a year, enough

Figure 14.2: Results of a field experiment to determine optimum fertiliser regime for establishment of a legume–grass sward. The site is an ameliorated lead–zinc mine waste in western Ireland (Jeffrey and Maybury 1981)

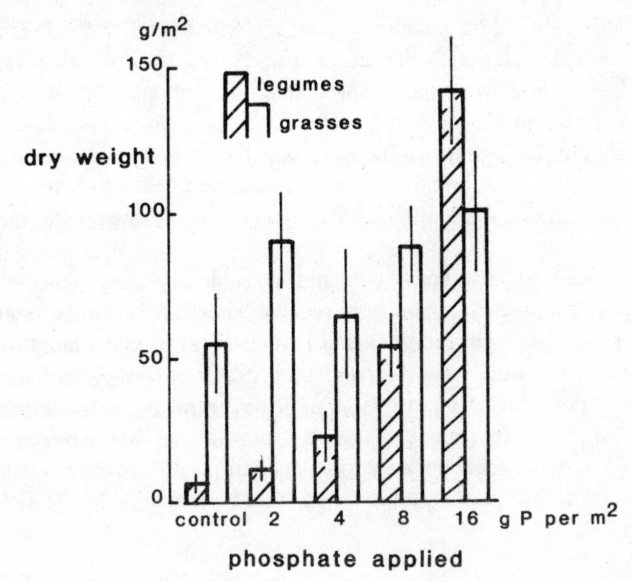

Figure 14.3: Growth of *Festuca rubra* seedlings in two sand culture experiments incorporating solid-phase metal sulphides (left, CuS; right, PbS). Note the difference in concentration producing a given growth inhibition

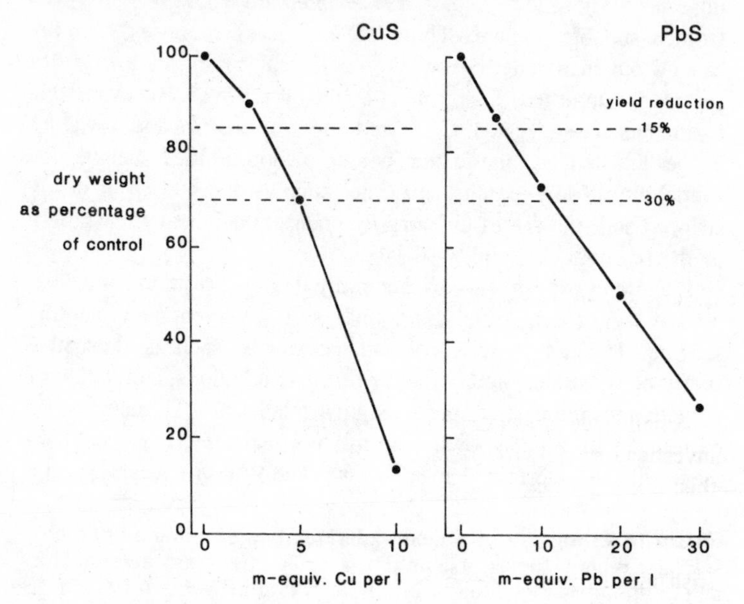

information should be gained to enable an optimum set of treatments to be applied with a high degree of confidence. All workers in this field know, however, that unpredicted interactions may lead to problems at a much later stage. Hence the need for monitoring.

In Figure 14.2, the results are presented of a field experiment which sought an optimum phosphate addition treatment for the establishment of a legume–grass sward on an ameliorated metal-rich waste.

In some circumstances it is not possible to carry out field experiments that give meaningful results. These include continuous waste-producing operations such as mine tailings, red mud from alumina production, or pulverised fuel ash from coal-fired power stations. For practical reasons it may be necessary to undertake much smaller experiments ranging from rather expensive pilot schemes to small-scale pot experiments. This procedure may narrow the range of practical options to be tested eventually at the field scale.

Figure 14.4: Design for 1 m^2 lysimeters for investigating nitrogen leaching loss and movement of toxic metal ions under experimental mining waste revegetation treatments

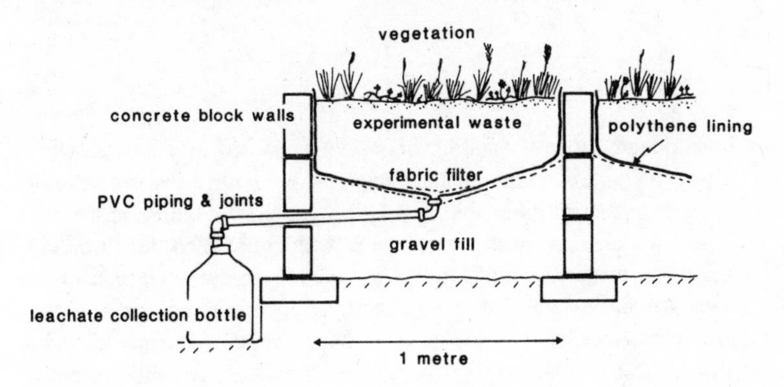

Small-scale experiments do have other virtues, in that they permit investigations not possible at the field scale. Two examples illustrate this:

(a) The testing of toxicity of individual minerals

Irish metal mining wastes have several toxic components, chiefly sulphides of copper, lead and zinc in a calcium carbonate matrix. Sand culture experiments have quantified the relative toxicity of base metal sulphides. This experimental system has also been used to screen plant species for growth and metal uptake (Figure 14.3). These experiments also show that the dilution of solid-phase sources of these metals plays an important part in the reduction of toxicity.

(b) The quantification of nitrogen cycle processes

As part of a study of all aspects of the nitrogen cycle as they relate to metal mining wastes, Levinge (1977) utilised a lysimeter system with 1 m^2 tanks (Figure 14.4). This system facilitated the collection of leached nitrogen, as well as permitting pilot-scale trials with legume–grass mixtures. Leaching of metal ions from this system was also studied (Jeffrey and Maybury 1981).

6. Site preparation

This stage entails the implementation of the optimum ameliorant scheme. It should be emphasised that sites are rarely homogeneous.

It is probable that variations may need to be applied. For example, some areas may require lime to raise pH, or land drainage, whereas others may require physical stabilisation.

7. Seeding

In some situations normal agronomic procedures may be followed, using conventional equipment. However, in many cases steep slopes or other accessibility problems may rule this out. Hydraulic seeding is now widely used in such situations. A slurry of seed, fertiliser and a binder, usually wood pulp, is sprayed over the surface using a high-pressure water pump. As well as overcoming accessibility problems, this technique permits a wide range of inoculants to be used. These can include *Rhizobium* or fungal cultures and other soil organisms. There is scope for a great deal of research into creating optimal inocula for hydraulic application.

8. Monitoring and maintenance

In this procedure a number of ecological techniques can be employed. These include assessment of sward botanical composition and elemental absorption, observations on chemical and biological activity in soil, and determination of herbivore activity. The extent of wind or water erosion may also be important. Unlike an agronomic operation, sheer production is not necessarily a prime consideration; more important is the establishment of ecosystem processes to a point where they are sustained with minimum management.

9. Additional planting

Overemphasis on the use of grasses and herbs may not be helpful, particularly in more arid climates. Woody species can be introduced early, and a mulch of pulverised vegetation can be used as a seed source. However, it is often useful to condition and stabilise a site by planting non-woody species and subsequently to introduce trees and shrubs. Nitrogen-fixing genera, such as *Alnus* (alder), *Hippophaë* (sea buckthorn) and woody legumes, may be of special value. There are close comparisons between this phase of land restoration and the ecology of pioneer species.

DEALING WITH BASE-METAL-CONTAMINATED SITES

Mining for metals, e.g. copper, lead, zinc and nickel, gives rise to three kinds of waste material: tailings, the finely ground residues of the 'gangue' mineral from which the ore has been largely extracted; waste ore, which contains a less than economically extractable concentration of ore minerals; and overburden, the non-ore-bearing material which must be removed to reach the ore. This will have a low toxic metal content and may be used to improve rehabilitation. The ratio of overburden to other wastes will be higher in opencast than in underground mines.

Two different approaches have been adopted in the literature towards revegetation of such wastes; to seek 'metal-tolerant' ecotypes; and to 'ameliorate' toxicity permitting growth of many species. The most sensible view of these approaches is to regard them as being complementary in a practical sense.

The phenomenon of metal tolerance is a fascinating topic with a large literature reviewed by Antonovics *et al.* (1971). Geological anomalies and mining wastes giving rise to elevated soil concentrations of copper, zinc, nickel, etc. may bear a sparse natural vegetation. If the clonal plant material is tested for capacity to grow in the face of normally toxic metal concentrations, a specific tolerance is demonstrable. A convenient screening test devised by Wilkins (1957), which entails measuring elongation of roots in solutions containing the appropriate metal ion, can be used to detect tolerance. Use of this test has shown a number of important aspects of tolerance:

(a) Tolerance is very specific to the metal ion selected against in the field.
(b) Cotolerance to several metals is virtually unknown.
(c) Few species possess tolerance. In Europe a number of grasses, e.g. *Festuca rubra, F. ovina* and *Agrostis stolonifera*, possess usefully tolerant ecotypes.
(d) Selection of a tolerant population may be rapid.
(e) Inheritance of tolerance is probably complex.
(f) Tolerant races may be commercially developed by selection. It is possible that in many situations a tolerant population can be readily selected from the pre-existing gene pool.

There are two vulnerable features of vegetation schemes that are heavily dependent on tolerant grass ecotypes:

(a) The provision for nitrogen nutrition may break down, especially as legumes are generally non-tolerant. This problem could be overcome if tolerant free-living or rhizosphere-associated nitrogen-fixing microbes could be isolated.
(b) Mixed toxicity conditions, e.g. toxic quantities of copper and lead, render tolerance ineffective because of the lack of cotolerance.

Knowledge of these limitations led us to use an 'ameliorative' approach towards contemporary mine wastes in Ireland (Jeffrey *et al.* 1975). This approach requires that toxicity be sufficiently reduced to permit establishment and sustained growth of suitable vegetation cover. Vegetation can be established, but more work needs to be done on the effectiveness of the decomposer food web. It is not clear if the rather depauperate set of soil organisms observed is merely a function of time or whether the metal toxicities are exerting a long-lasting effect. Another unanswered question concerns the establishment of deep-rooted shrubs and trees. Will the exploration of deeper horizons by roots and mycorrhizas give rise to delayed toxicities as toxic metal ions are encountered?

Mechanisms for the reduction of metal toxicity

(a) Dilution of sources of toxic metal ions

The solid-phase sources of toxic metal in metal mining wastes are usually the sulphides or carbonates of the element. Availability is a function of the length of the diffusion pathway from mineral particle to absorbing surface. Dilution tends to increase this path length. Responses to differing concentrations of solid-phase sources illustrate this. An example is shown for lead in Figure 2.4, but the effect also applies to copper and zinc sulphides.

(b) The addition of organic matter

Polyvalent metal ions are bound both reversibly on cation exchange site and almost irreversibly, by liganding bonds, to organic surfaces. Addition of small quantities of peat, paper pulp or organic wastes appears to provide alternative absorption sites for ions such as Cu^{2+}. Copper is a comparatively toxic element (Figure 14.5), but sufficient amelioration can be achieved using organic matter to permit growth of sensitive legumes. A technique that needs exploring is the growth of an annual cover crop to provide organic matter.

Figure 14.5: Growth of *Festuca rubra* on a copper sulphide sand culture with added organic matter (Jeffrey, Maybury and Levinge 1975)

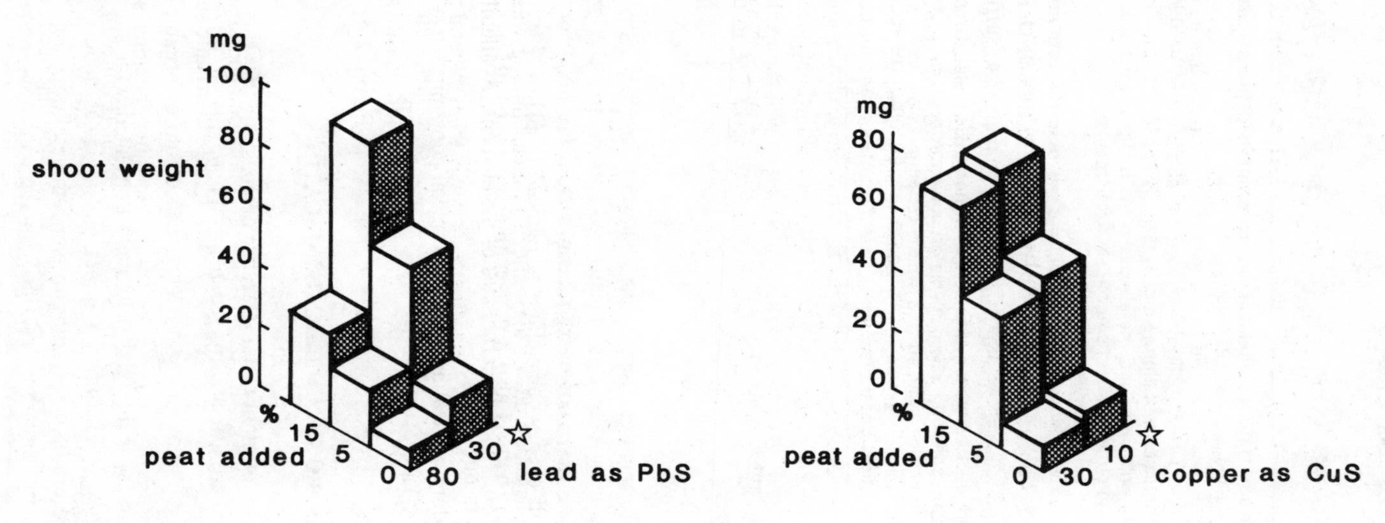

Table 14.2: Procedure for dealing with base-metal-contaminated substrates

(1) Analyse substrate, determining total concentration, mineral forms and possibly a measure of availability.
(2) Determine by simple field or pot experiments if the substrate is toxic to a range of test species, e.g. legumes and grasses. Would metal-tolerant species be helpful?
(3) If growth inhibition and metal uptake are marked, try the following ameliorative measures in small-scale experiments:
 (a) Raise pH by liming.
 (b) Dilute with non-toxic material, e.g. overburden, other wastes.
 (c) Add a source of organic matter, e.g. paper pulp, peat, sewage sludge
(4) If toxicity persists, quantify the problem of covering waste with a relatively thick layer, e.g. 25–30 cm, of non-toxic material. Test on a pilot scale, checking also possible groundwater pollution problems.
(5) If toxicity can be reduced to permit establishment of plant cover, are there problems in scaling up the ameliorative treatment? Check with pilot-scale field experiment.

Suitable tolerant species may pave the way for a succession of less tolerant types.

(c) Adjustment of pH

Most metal toxicities are greatest at low pH and if a siliceous or pyrite-containing gangue is encountered, pH control is a priority. The substrate may, however, remain sufficiently toxic to prevent plant growth even when acidity is reduced to pH > 7.0 by lime application. There is also a danger that superficial pH adjustment will revert in due course.

A procedure for dealing with a base-metal-contaminated substrate is set out in Table 14.2.

PLANTS AS BIOGEOCHEMICAL INDICATORS

The examples of plant metal tolerance demonstrate that species and communities are affected by geochemical anomalies in three major ways:

(a) They are sensitive and negatively correlated with anomalies. This may mean either a direct physiological toxic effect or reduced production leading to competitive elimination. This is the general situation described for the serpentine anomaly in

Chapter 19, when outcrops are encircled by non-serpentine vegetation.

(b) They are tolerant and survive best under low-competition conditions. This may give rise to either simplified vegetation or a radically different vegetation type. A good example is the zinc vegetation of western Europe, which includes a series of tolerant species and ecotypes: *Viola calaminaria, Armeria halleri, Thlaspi alpestre* var. *calaminare, Silene cucubalus* and *Agrostis tenuis. Minuartia verna* (Caryophyllaceae) in lowland sites is a good indicator of soil lead, but not necessarily in its more montane range.

(c) Species are both tolerant to and accumulate the anomalous metal. In this situation the physiological basis for tolerance includes the capacity to sequestrate ions in tissue compartments separated from cellular metabolism (see Peterson 1971; Woolhouse 1983). An often quoted example is the labiate *Becium homblei* of central Africa which accumulates copper to concentrations of 3000 μg Cu g d.w. How the capacity to accumulate evolves within particular taxa is not properly resolved, but several species within particular families seem to possess the characteristics, e.g. three *Alyssum* species (Cruciferae) are known as nickel accumulators (cf. the serpentine example, Chapter 19); a number of *Silene* taxa (Caryophyllaceae) accumulate cobalt, tin, zinc and copper. A more systematic view might provide pointers to hitherto unsuspected accumulator species.

(d) The general vegetation composition may not alter in response to a low level of anomaly, but elemental concentrations of particular organs may yield valuable information. In this situation a common species is desirable, with an organ for analysis in which the seasonal variation does not complicate interpretation. Careful surveys are needed to isolate suitable test systems. The bark of trees, the long-lived leaves of evergreens, and young stem tissues of deciduous trees have all been used for this type of survey. Forest humus is also seen to give more valuable information on bedrock type than the underlying soil.

These effects have obvious potential in prospecting for mineral deposits, and may be regarded as complementary to geological survey techniques such as stream-bed sediment monitoring, magnetic surveys and detailed investigations of geological structure (Brooks 1979).

Changes in vegetation type brought about by geochemical factors may be detected either in ground survey or by remote sensing procedures. Properly processed multispectral images from satellites may present a clearer image of subtle vegetation changes than conventional aerial photographs.

DEVELOPMENT OF HYPERACIDITY IN MINE WASTES

The mineral iron pyrite, FeS_2, is sometimes associated with other more valuable minerals including coal, lead, zinc, nickel, gold and copper ores and barytes ($BaSO_4$). It may thus become a component of mining wastes and ore processing residues. It is also encountered in the field when coastal sediments are reclaimed (see Chapter 18). Substrates known to contain pyrite are notorious for their tendency to generate acidity, with acid substrates and the production of acid waters as a consequence. A series of only partially understood chemical and biological reactions leading to the oxidation of pyrite is summarised by this pair of equations:

$$\underset{\text{Iron pyrite}}{2FeS_2} + 2H_2O + 7O_2 \rightarrow \underset{\text{Ferrous iron}}{2Fe^{2+}} + 4H^+ + 4SO_4^{2-}$$

The further oxidation of ferrous iron gives rise to yet more acidity:

$$4Fe^{2+} + 10H_2O + O_2 \rightarrow 4Fe(OH)_3 + 8H^+$$

In soil the acidity generated leads to a series of effects readily predictable from soil chemistry as discussed in Part II, viz. reduced nutrient availability; decreased biological activity including mineralisation and nitrogen fixation; and mobilisation of toxic elements including Fe^{2+}, Mn^{2+}, Al^{3+} and any other toxic metals if present. These problems are also transferred to fresh waters, with the additional difficulties caused by iron and aluminium salts and their conversion to flocs of hydroxyl compounds. Massive poisoning of rivers with copper and zinc ions following pyrite oxidation is well known.

The overall production of acidity will proceed slowly as a purely chemical reaction set, but acid generation is at least twenty times more rapid with active microbial populations. Ultimate control of acid production, not totally achievable yet, requires a better insight into the chemistry and microbiology of the partial reactions. These are part of a reaction cycle which must be broken.

Acid production may be separated into a series of interlinked partial reactions, after Van Breeman (1973) and Tubridy (1983).

(1) Initial attack of pyrite:

$$FeS_2 + 1/2O_2 + 2H^+ \rightarrow Fe^{2+} + 2S^o + H_2O$$

In the field, reaction rate is strongly controlled by the grain size of the mineral (Caruccio 1975). This means that it is difficult to measure the acid-producing potential of a particular waste. It is not clear if microorganisms are directly involved in this reaction, which will proceed if the pH is less than 4. At a higher pH, Fe^{3+} is formed and ferric salts precipitate. More critical work is needed using pyrite as a substrate in laboratory studies.

(2) Pyrite attack by ferric ions:

$$FeS_2 + 2Fe^3 \rightarrow 3Fe^{2+} + 2S^o$$

This is another inorganic mode of attack, again generating elemental sulphur.

(3) Oxidation of iron by *Thiobacillus ferrooxidans*:

$$Fe^{2+} + ¼O_2 + H^+ \rightarrow Fe^{3+} + ½H_2O$$

This could be a key rate-controlling step, with the bacterial oxidation rate increased over the chemical rate by 10^6 times. A series of chemautotrophic organisms are capable of utilising the reaction, but *T. ferrooxidans* is the most commonly isolated in the acid range (pH 3.5–2.0).

(4) Microbial oxidation of elemental sulphur:

$$2S^o + 12Fe^{3+} + 8H_2O \rightarrow 12Fe^{2+} + 2SO_4^{2-} + 16H^+$$

Elemental sulphur is commonly observed on the surface of pyrite grains together with chemautotrophic sulphur bacteria. Both *Thiobacillus thiooxidans* and *T. ferrooxidans* can utilise S^o as a substrate to produce large quantities of H^+ ions. (Generation of acidity by this reaction can be used to produce a low-technology phosphate fertiliser for tropical conditions. Rock phosphate and sulphur are pelletised with a suitable *Thiobacillus*. The acid produced liberates available phosphate when wetted in the soil. The reaction proceeds best under high soil temperatures.)

Control of acid formation

By noting the reactants in production of acid, especially oxygen, water, FeS_2 and hydrogen ions, it is possible to consider hypothetical schemes for amelioration.

(a) *pH control by liming*. This has been much practised, and is frequently the most disappointing control measure. The capacity to produce acid is very high when compared with the economic dressings of limestone that may be incorporated merely in the surface.

(b) *Reduction in partial pressure of oxygen, and oxygen diffusion rate*. This tactic may be of value in acid-producing mine workings but to date has not been successfully applied to waste dumps. Heterotroph activity can be stimulated by the addition of organic wastes such as sewage sludge or animal manure. However, this does not seem to be sufficiently intense under field conditions.

(c) *Control of water*. Two approaches can be applied:

   (i) To leach dumps with the objective of deliberately removing pyrite from the upper horizons, then treating residual acidity with lime. This is more likely to be effective in an arid climate with superficial wetting normally.

   (ii) Restricting the penetration of water into dumps. This approach is unconsciously followed when low-toxicity coal strip-mining wastes are prepared for agricultural production. Careful compaction with rollers to avoid subsidence prevents rain water from reaching deeper horizons. Diverting any surface watercourses away from pyritised wastes or bedrock seems to be an important procedure.

(d) *Control of pyrite*. In the practice of mining, pyrite may be thought of as either useless or as a saleable by-product. The mineral can be used as a raw material in sulphuric acid production. Furthermore, the whole process of acid production may be utilised for the leaching of low grades of ore, especially copper ores. However, it is rare that removal of pyrite will be a widely available solution. It may be possible in some cases either to concentrate pyritic wastes deep in a dump or to disperse small quantities through calcareous wastes. The problem of dealing with inherited acid-producing conditions generally rules out control of pyrite.

Contemporary pyrite-containing wastes

All the above tactics can be considered when disposing of currently produced wastes. The cost of each may be assessed in the light of water quality and site restoration costs. Isolating pyrite from air and water would seem to be key priorities in disposal strategy. Calcareous wastes, rather than bought-in limestone, may also have a role in buffering substrate pH.

15

Two aspects of forest mineral-nutrient economy

Forest ecosystems contain some 90% of the Earth's biomass and cover about 40% of its land area. Timber is a key natural resource as fuel wood and as a construction material. Study of forest ecosystems is of vital human importance in securing sustainable yields of energy and materials and understanding the impact of deforestation of watersheds. Deforestation arising from timber cropping, or to secure land for agricultural crop production, may be one of the most profound changes that man can inflict on the surface of the planet. The consequences of deforestation need to be spelled out to political leaders world-wide.

For various reasons of scale and longevity, forests offer ecology a window on processes not readily observable elsewhere. The coupling of plant cover with the hydrological cycle in particular may be seen and measured. The two aspects of woodland plant–soil relationships illustrated are both linked to this theme. First, mineral cycling within two contrasting temperate forest types will be illustrated. This is followed by a sketch of studies at the watershed level as illustrated by the renowned Hubbard Brook project.

MINERAL CYCLING IN FOREST TREES

There has been a 1000-year-old tradition of managing tree crops in Europe. This has utilised many management systems including coppicing, selective felling, thinning and clear felling. Practical mensuration techniques have developed in order to compare yields of sites and even to tax current or future income from tree crops. Forest mensuration is based on allometric techniques in which a few simple measurements, e.g. tree girth at breast height, can be converted to timber yield (Newbould 1967). Ecosystem measurements of forests have evolved from this approach. The first studies carried out were of even-aged stands of forestry species. These employed:

(a) Determination of biomass by direct and allometric methods. Determination of underground production and production by understorey species was a distinct break from pragmatic forestry studies.
(b) Chemical analysis of the various plant parts.

The results of biomass and tissue analysis enabled a view to be obtained of the utilisation of mineral resources by the woodland, balancing nutrient uptake with loss in leaf fall and other shed parts. These early studies demonstrate that biomass and nutrient content for a particular woodland type reaches a characteristic asymptote with time. Thus a hierarchy of forest types can be established according to ceiling biomass when production is matched by consumption of grazers and decomposers.

It has also been realised for 50 years that the ionic quality of rainfall is altered by passage through a forest canopy. This flow of ions can be compared with that involved in the growth of the woodland.

Rainfall impinging on a forest canopy can have several fates:

(a) evaporation, any ions being deposited on the leaf surface;
(b) throughfall;
(c) leaf wash ((b) and (c) are difficult to separate, but it is important to recognise that some ions may be removed from the canopy);
(d) stem flow or precipitation intercepted by the architecture of the trees and flowing as water films over the bark-covered surfaces rather than falling as drops. This needs to be specifically intercepted by specially designed and installed cups.

Because of the intensity of effort required, it is seldom that both tissue nutrient turnover and the interaction with precipitation are studied together. The two examples are selected from the compendium of IBP data edited by Reichle (1981).

Liriodendron woodland is a vegetation type characteristic of eastern North America. *Liriodendron tulipifera* (tulip tree) is virtually the largest deciduous species in the USA, but its behaviour is not very different from that of other temperate deciduous species, e.g. *Quercus* (oak) or *Fagus* (beech) (Figure 15.1). As a contrast,

Figure 15.1: Annual budget for potassium and nitrogen in deciduous woodland dominated by *Liriodendron tulipifera*

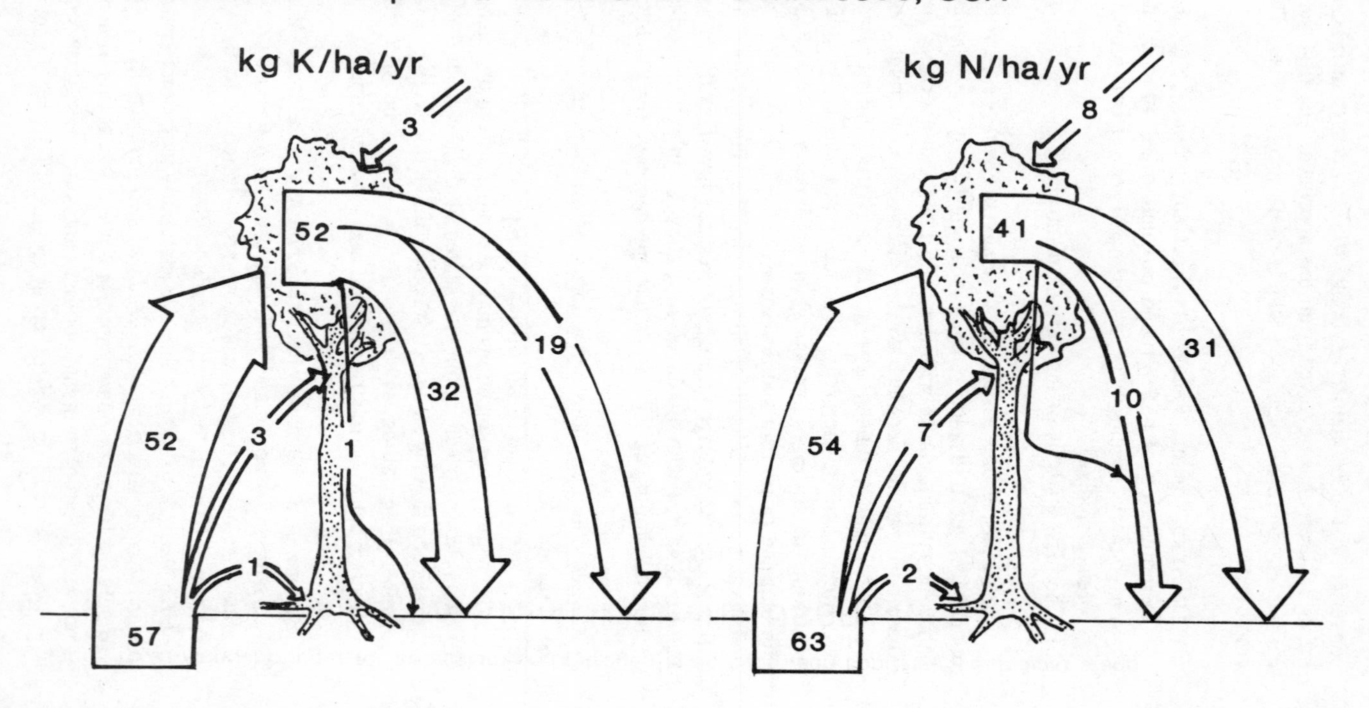

Figure 15.2: Annual budget for potassium and nitrogen in an evergreen conifer (*Picea abies*) stand

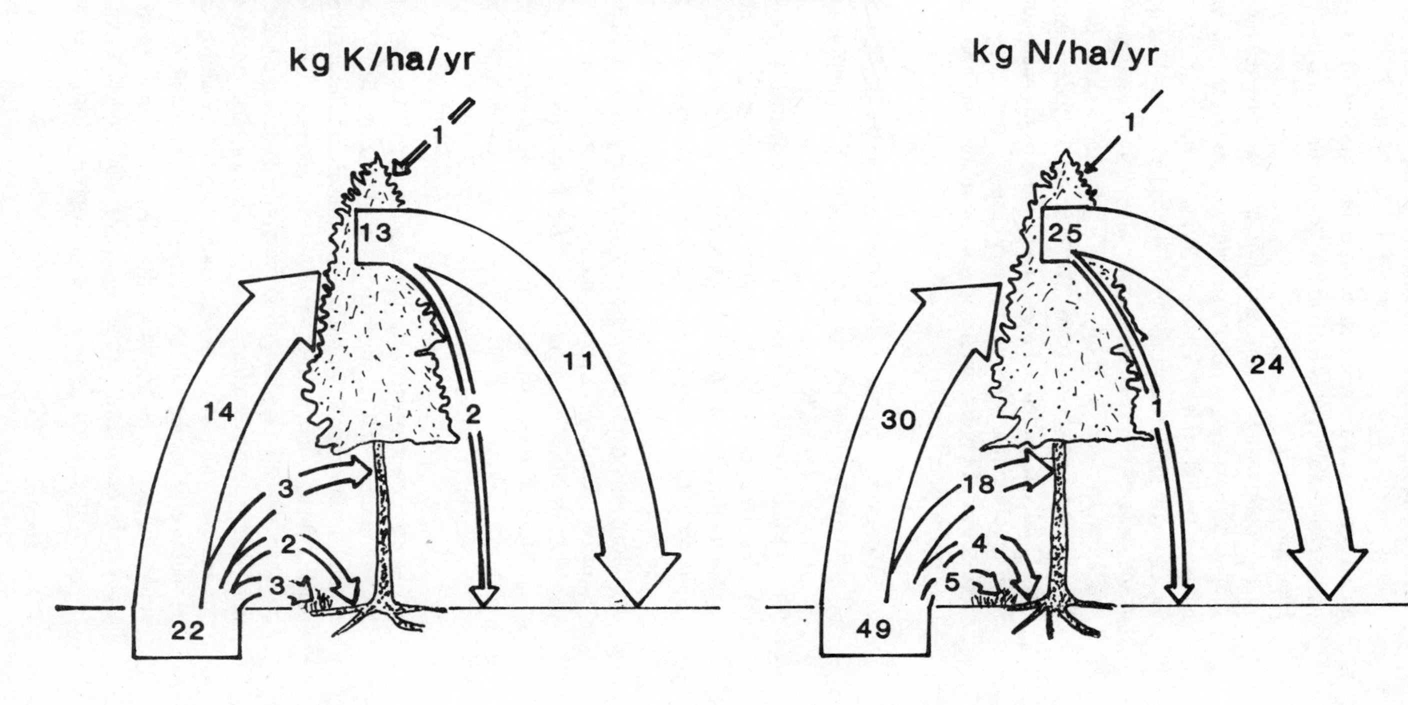

a data set from a *Picea abies* stand in the USSR is illustrated. This type is probably typical of much boreal forest in the Northern Hemisphere (Figure 15.2).

Two contrasting elements are chosen, potassium, the biologically simple monovalent cation, and nitrogen, with all its biological complexity and multiple chemical forms. In the diagrams, width of the arrow is proportional to the total uptake of the element, shown at the base of the upward-directed arrows (left-hand side). These symbolise transfer to canopy, trunk, roots and ground flora in the *Picea* example. Loss from the canopy includes stem flow, throughfall (combined in *Picea*) and leaf loss (right-hand side). Input from precipitation is shown in each case.

A number of points with widespread validity may be made from the comparisons:

(a) Turnover is higher for both elements in the deciduous broad-leaf system. Conversely, less of the absorbed ions are returned in the *Picea* system.
(b) The leaves of the canopy are the largest sink for absorbed ions.
(c) Leaf wash can remove substantial quantities of potassium: 61% of total return in *Liriodendron* and 15% in *Picea*. Less nitrogen is removed by this process: 24% in *Liriodendron* but only 4% in *Picea*.
(d) More nitrogen is assimilated by the canopy or recirculated internally. Virtually all potassium absorbed is returned.
(e) The variation in inputs from the atmosphere is probably a difference in: distance from the ocean (potassium); prevailing wind; and pollution (nitrogen). In woodland communities these inputs are relatively small.

We can now appreciate that the exploitation of forests must make use of the elemental turnover data revealed by studies of this kind. When this approach is applied to tropical rainforest, the results indicate that much of the total mineral nutrient resource of the ecosystem is actively engaged in the cycle. Thus large-scale vegetation destruction to clear land for agriculture leads to very considerable losses of the nutrient capital. Viability of such agriculture is suspect in the long term. When combined with the extinction of species and the displacement of indigenous human groups, the ecological case against particular modes of exploitation is very strong. Ecology here has great political implications.

AN EXPERIMENTAL STUDY OF SMALL AFFORESTED CATCHMENTS AT HUBBARD BROOK, NEW HAMPSHIRE

The three scientists who initiated this programme of research, Gene E. Likens (ecologist), F. Herbert Bornman (forester) and Robert S. Pierce (hydrologist), were stimulated by the value of experimental manipulation of lake ecosystems. They recognised that a set of similar small watershed ecosystems in the Hubbard Brook Experimental Forest represented an opportunity for experimental manipulation of broad-leaved forest.

This study directly addresses the question, 'What are the geochemical effects of forests on catchments?' The study is remarkable in its original conception and initiation in 1963, in the thorough and consistent monitoring over a 20-year period in cooperation with the USDA Forest Service and in the allocation of funding over the period. The study has been described in a prolific literature now brought together as a bibliography (P.C. Likens 1984). Most convenient access to the literature is via a book (G.E. Likens, Borman, Pierce, Eaton and Johnson, 1977) or, more recently, a personal commentary by G.E. Likens (1985).

An experimental study of deforestation

It is important to recognise that the unit of study is not the biological or pedological details of the forest ecosystem but the hydrologic inputs and outputs (Figure 15.3). The deforestation experiment is one of the most central to soil–plant relationships and it is worth describing briefly. One watershed was deforested in the winter of 1965–66 by careful felling, leaving fallen timber in place. Herbicides were applied for two seasons, keeping the watershed vegetation-free. Natural recovery was then permitted.

From 1963 to the present time, two types of monitoring have been applied to the deforested area and an undisturbed reference watershed.

(a) A network of rain gauges was used to determine total precipitation as rain or snow, and to measure the concentration of major ions, e.g. sources of nitrogen, sulphate, calcium, magnesium, potassium, sodium and phosphate. The product of volume and concentration, expressed as kilograms of ion per hectare per year, represented ion input.

Figure 15.3: Diagram of the unit of study in the Hubbard Brook watershed research programme

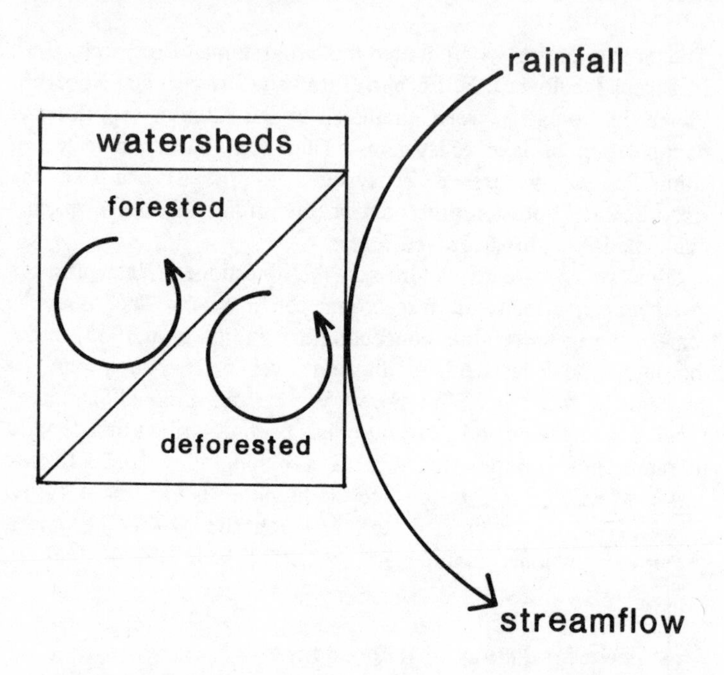

(b) The primary stream leaving each catchment was fitted with a 'vee-notch' weir to measure flow rate. Concentration of ions was determined by analysing water samples collected from gauging stations. Again, the output values could be expressed in terms of the watershed area.

From a comparison of the input and output values for a particular ion, a mass balance budget could be calculated. A highly schematic composite picture of this budget, applicable to nitrate, potassium and magnesium, is presented in Figure 15.4.

In the undisturbed watershed, the nitrogen budget appears balanced, with a low level of loss detectable for potassium and magnesium. Immediately after deforestation, massive loss of all three ions is detected, with subsequent recovery to values close to the control. The nitrate budget recovers to slightly above control whereas the potassium budget is below control. The budgetary losses can be resolved into a series of components:

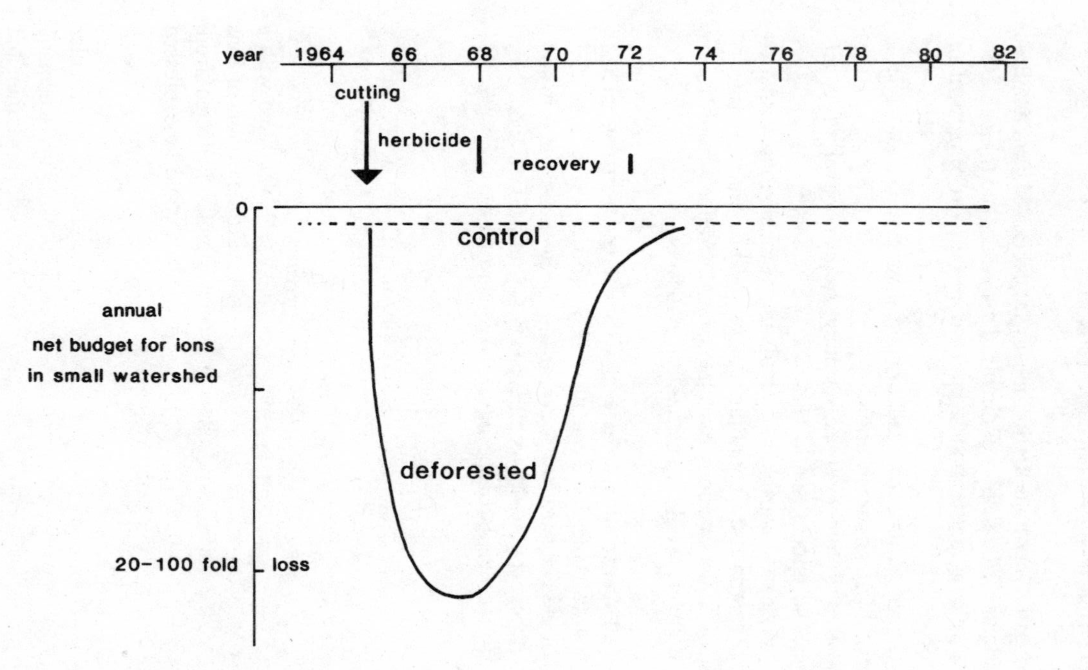

Figure 15.4: Schematic and simplified budget for ions in a deforested watershed

(1) Reduced evapotranspiration means (a) increased flow volume and increased transport of dissolved materials; and (b) increased flow velocity and a modest increase in transport of particulate materials.
(2) The large volume of dead organic matter is a pool of ions vulnerable to mineralisation.
(3) Higher soil temperatures resulting from removal of cover accelerate mineralisation.
(4) The vegetation capable of assimilating available ions was lacking. There is also a possibility that removal of control or inhibition of mineralisation, by allelopathic secretions from living plant cover, would accelerate mineralisation.

How general are these statements of the coupled hydrological and biogeochemical roles of vegetation in a landscape? The special features of the Hubbard Brook system are the juvenile landform and a mature forest cover with small capacity for nitrogen fixation. Not known are the details of the internal cycling of ions.

16

Australian heathlands and other nutrient-poor terrestrial ecosystems

Australian heathlands have a well documented plant ecology which illustrates the value of a long series of interlocking studies. These have considered the biogeography and derivation of the distinctive flora, its associated soils and their low fertility, the role of fire, water relationships, productivity and ecophysiology. The available information raises questions regarding the uniqueness or otherwise of these shrub-dominated formations. Six suites of species are recognised in the whole continent (Specht 1981), but most of the work referred to was carried out in southeastern Australia.

Australian heathlands are dominated by shrubs usually less than 2 m high, with small to medium leaves which are tough and leathery (sclerophyllous). Trees are typically absent, but integradations with woodland or savannah grassland are encountered. Heathland shrubs may persist as the understorey of adjacent eucalypt woodland. Usually three or four species codominate. Table 16.1 lists the key families contributing to heath vegetation from subhumid southern and southeastern Australia (Specht 1981). This collection of taxa may be interpreted as a relic of a tropical flora isolated by aridity in the Pleistocene–Recent period. Treeless heathlands are regarded as climax communities rather than transitional stages to woodland. Grasses are also absent except as transitional or ephemeral species after fire.

Table 16.1: Taxa contributing to heathland vegetation in subhumid southern and southeastern Australia (from Specht 1981)

Family	Genera	Habit
Monocots		
Anthericaceae	5	Lily-like geophytes
Cyperaceae	4	Sclerophyllous evergreen graminoids
Orchidaceae	43	Geophytes
Poaceae	5	Seasonal grasses, usually successional
Restionaceae	5	Sclerophyllous graminoids
Dicots		
Asteraceae	10	Short-lived seasonal or successional daisies
Epacridaceae	9	Sclerophyllous subshrubs, cf. Ericaceae
Euphorbiaceae	5	Rare or transient herbs
Fabaceae	16	Nodulated nitrogen-fixing legumes, usually perennial
Goodeniaceae	4	Annual or perennial herbs or subshrubs
Myrtaceae	9	Sclerophyllous shrubs
Proteaceae	8	Sclerophyllous shrubs
Rutaceae	8	Sclerophyllous shrubs
Subtotal	101	
Others[a]	17	
Total	118	

[a] Individual genera of importance not included in families noted above: Xanthorrhoreaceae, 1 genus: sclerophyllous shrub 'grass trees'; Casuarinaceae, 1 genus: sclerophyllous shrub with *Frankia*-type nodules; Droseraceae, 1 genus: carnivorous plants, usually geophytes; Mimosaceae, 1 genus: *Acacia* species.

SOILS

Patches of heathland occur in Australia, in situations ranging from tropical to temperate and sea level to alpine. The soil type is inevitably infertile even though derived from a wide variety of parent materials. The heathland soils of South Australia and Victoria, for example, are deep sands, blown from ancient decalcified coastal dunes and podsolised in the recent period. In the alpine zone of southeastern Australia a heath type occurs on deep *Sphagnum* peats.

The idea is fairly well established that the boundaries of many plant communities in Australia are delimited by the phosphate content of soil. A picture is painted by Beadle (1966) of a situation in which xeromorphic genera decline and rainforest genera increase with increasing soil phosphorus. Heath sites represent one end of this continuum (Figure 16.1). In general, heathland mineral soils

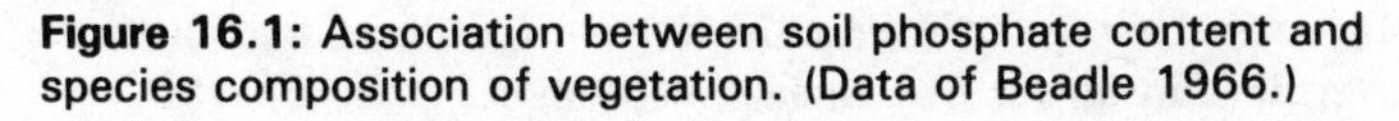

Figure 16.1: Association between soil phosphate content and species composition of vegetation. (Data of Beadle 1966.)

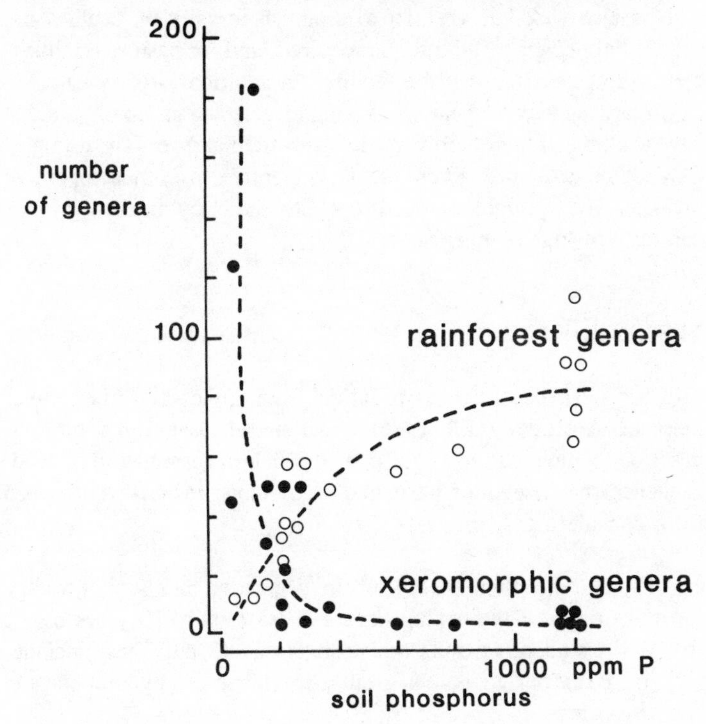

have total phosphate-P contents of less than 75 ppm and nitrogen contents of less than 0.1%.

THE EFFECT OF FERTILISERS

A series of fertiliser addition experiments have now been carried out on heathland vegetation, the first key experiment being that of Specht (1963). Response is to added phosphorus in the first instance in all cases, with the dominant species increasing in biomass by two- to three-fold. More mesophytic growth forms may be adopted by dominants, and the leaf area index increases. In low rainfall seasons, lethal water stress may kill 50% of fertiliser-stimulated dominants (Specht and Jones 1971). Invasion of gaps between shrubby dominants by herbs usually occurs on fertilised plots. These are frequently non-native introduced species. There is evidence that the long-term effect of fertiliser addition is towards the extinction of the heath flora in favour of savannah vegetation.

It is also well known that, to establish forestry on heath soils, substantial phosphate addition is required, and for pasture establishment trace elements may be required in addition to phosphate.

So far, the view of heath is of a highly selected group of taxa that have acclimated to both low nutrient substrate and periodic drought. The set of taxa in a given site is vulnerable to both competitive invasion and drought on fertilised sites and, by inference, sites naturally of higher nutrient status.

RESPONSE TO FIRE

Fire has been described as a natural environmental variable over most of Australia (Gill 1975). Various adaptive traits may be described which lead to vegetative survival or germination of seed. Evidence that fire is of particular importance in heath is derived from considering:

(a) The extreme flammability of the vegetation and the knowledge that present-day burning occurs at least every 50 years or so.
(b) Various adaptive traits are strongly developed. These include:
  (i) Buds buried below ground in rhizomes, lignotubers and corms.
  (ii) Seeds that are dispersed by fire, e.g. Proteaceae, Casuarinaceae, Myrtaceae: the woody fruit follicles are long lived, storing seed for several years. (To obtain seed for experimental work from *Banksia* species (Proteaceae) it is convenient to heat the whole follicle for some minutes over a bunsen burner. The lips of this complex follicle then open to release the seeds with their slender wing.)
  (iii) Fire-resistant stems. These are not particularly common in heath, but *Xanthorrhoea* is a spectacular example. It is a member of the Liliales, and the 'trunk' of leaf bases and exuded resin protects buried buds.
  (iv) Fire-stimulated flowering. *Xanthorrhoea* is again a good example as whole populations respond by producing large flower spikes on long scapes up to 1 m high. Flowering is otherwise rare. At least one heath orchid is known to behave in a similar fashion.
(c) The succession of vegetation development in heath following

fire. Although the stems of most woody species are vulnerable to fire, many species regenerate rapidly from protected buds. The first signs of regeneration are sprouting stems from rootstocks of woody species. In the bare ground between these shoots there is a flush of forbs and grasses which may recur for a few seasons. A large number of perennial species, including the dominants, also germinate in the first few months after fire. These shorter-lived species eventually mature and die, usually by the tenth year after fire. Long-lived dominants regenerating from seed tend to be suppressed until this time, when their biomass increases to dominate the biomass of the stand at about 25 years. Species numbers decline and are minimal in the oldest known stands of heath, approaching 50 years.

PHOSPHATE ECONOMY

If it is accepted that all the field evidence points towards accommodating to an ultra-low-phosphate environment, then there are questions to be answered at the physiological level. It must be emphasised that neither introduced agricultural or forestry crops nor other forms of native vegetation can exploit the heathland substrate.

There are at present three topic areas which offer partial explanations of the survival strategy in the face of low phosphorus:

(a) the acceptance of fire;
(b) the scale, location and effectiveness of root systems in assimilating phosphate;
(c) the physiological manipulation of phosphates.

The acceptance of fire

The biological evidence of fire being a normal part of the heathland environment must be combined with what is known of the chemical effects of fire discussed in Chapter 6. In particular, it is seen from the pyric succession that much nutrient capital is immobilised in the standing crop after 25–50 years. It is clear that phosphate in particular is remobilised by fire. The regular occurrence of a flush of invading forbs for a few seasons after fire strongly resembles the effect of fertiliser addition. It is not clearly known, however, if phosphate losses occur as demonstrated for burned European heath (Gimingham 1972).

The scale, location and effectiveness of root systems

High underground biomass values are recorded for heath vegetation. For example, nine-year-old heath on deep sand at Keith, South Australia, with above-ground biomass of 7 tonnes per hectare, had a below-ground biomass of 58.5 tonnes per hectare (Groves and Specht 1965). Nearly 70% of this biomass is within the upper 15 cm of the soil and 90% in the top 30 cm. Some roots are found down to 1.5 m, however, a reminder that severe summer droughts may occur in some years.

The root biomass in this case is a complex mixture of types. The three dominant species are:

— *Casuarina pusilla*, a dwarf species which dominates for the first 15 years or so after fire. Its roots have the woody nitrogen-fixing nodules typical of the family, with a *Frankia* symbiont.

— *Leptospermum myrsinoides* (Myrtaceae) has a finely divided root system branching from a few major woody roots. The fine roots of both *Casuarina* and *Leptospermum* are presumed to be mycotrophic.

— *Banksia ornata* (Proteaceae) is the species that eventually dominates this formation, with a 50% contribution to stand biomass after 25 years. Its roots are unusual in that they are non-mycorrhizal (Purnell 1960); and that two morphologically distinct root types are produced, a closely branched system with a dense covering of long-lived root hairs very close to the surface (so-called 'proteoid' roots) and deeper roots that are less branched and with a conventional morphology (non-proteoid roots). The proteoid root mat has two interesting properties in that it forms a large plate-like mat surrounding each *Banksia* bush with a diameter of some metres. This may represent a macro-scale nutrient trapping surface. At the micro-scale, phosphate uptake rate per unit dry weight is approximately twice that of non-proteoid roots (Jeffrey 1967), as judged by ^{32}P absorption from solution. This effect is probably related simply to area of absorbing surface.

The root and rhizome systems of evergreen graminoids, in particular the Cyperaceae, e.g. *Lepidosperma* and Restionaceae, e.g. *Hypolaena*, are also non-mycorrhizal.

Mention should be made of the geophytes of the Liliales and Orchidaceae. Their efforts seem directed towards mycotrophy, nutrient storage and dispersal of fine seed. Another geophyte is the carnivorous plant *Drosera whittakeri*, whose bulbs are often embedded in the proteoid root mat of *Banksia ornata*. It is suggested that this and other carnivorous species have a phosphate supply independent of the soil (see Chapter 4).

The roots of heath vegetation seem well deployed to absorb any phosphate released by decomposition or entering as dust, ash or animal remains. There is, however, much scope for exploring the seasonal activity of the soil–root system with respect to phosphate and nitrogen, microbes and the soil solid phase.

The physiological manipulation of phosphate

Heath ecosystems have retained an unusual summer growth rhythm, which is interpreted as a token of their tropical origins. Growth begins rather suddenly with shoot elongation commencing in November and continuing through the summer. This conspicuous elongation is followed by fall of older leaves. Growth terminates in April, but leaf shedding extends into winter. Root growth seems to continue, with biomass low at the beginning of summer, doubling in the course of the shoot elongation season and reaching maximum values in the autumn.

Throughout this rhythm of growth, changes in phosphate turnover may be discerned. This study is not complete, and the best data available are from *Banksia ornata*. As the growing season starts, there is a mobilisation of phosphate which is detectable as a rise in trichloroacetic acid (TCA) soluble phosphate. This fraction normally contains orthophosphate and sugar phosphates. It declines as elongation ceases. As leaves senesce, phosphate is systematically scavenged from them. A small quantity only remains at abscission (Groves 1964).

^{32}P uptake studies have shown that phosphate absorption can occur throughout the year rather than in the growing season only. Newly absorbed phosphate typically remains in the root system in the short term, stored as long-chain polyphosphate (Figure 2.9) (Jeffrey 1964). This is an unusual feature in higher plants, phosphate usually being stored as phytin. Polyphosphate synthesis is possible in at least five other heath species, *Banksia serrata* and *Hakea ulicina* (both Proteaceae), *Casuarina pusilla* (Casuarinaceae), *Hypolaena fastigiata* (Restionaceae) and *Lepidosperma concavum* (Cyperaceae). All are important species in southeastern Australia (Jeffrey 1968).

The field significance of polyphosphate storage needs further exploration. Some evidence exists which shows the capacity of

Figure 16.2: Growth record through the summer period of *Banksia ornata* plants in water culture. A set treated for 24 h with 1 ppm phosphate-P is compared with a no-phosphate control

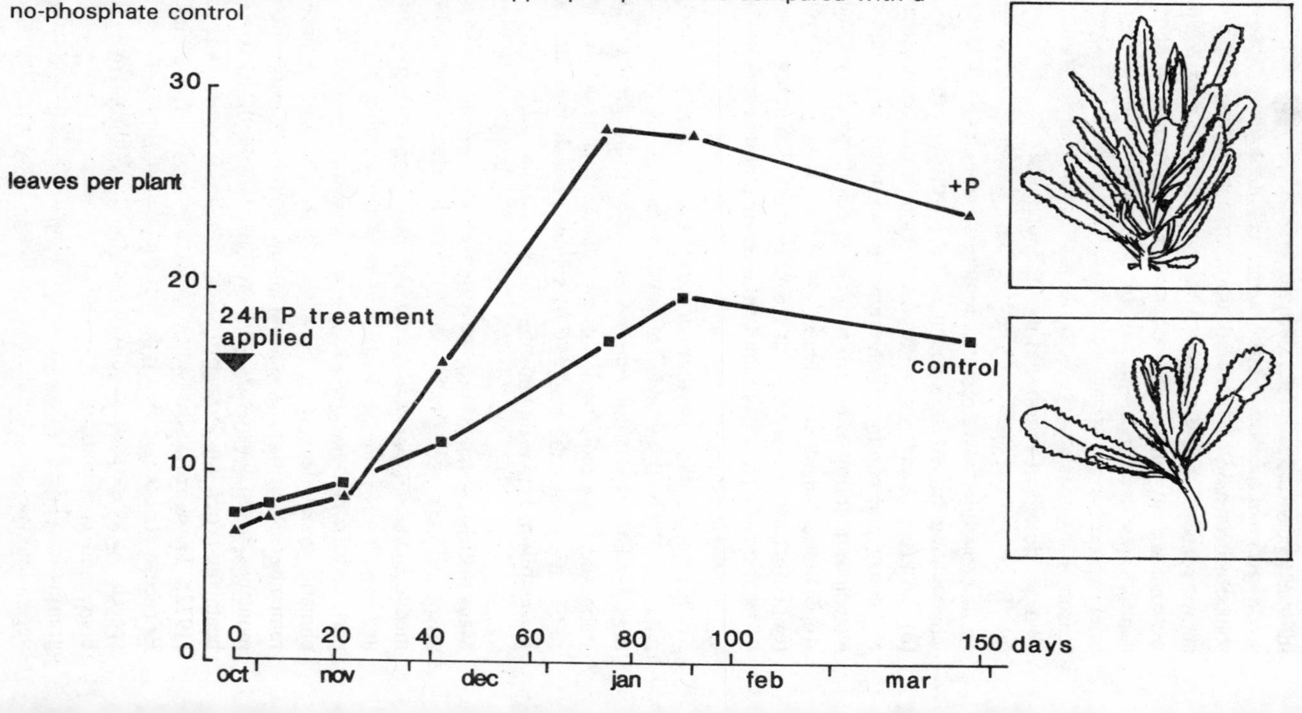

phosphate absorbed in the period before the growing season to influence growth. Half of an identical group of water-culture-grown seedling *Banksia ornata* plants were treated for 24 h in a culture solution containing 1 ppm phosphate as orthophosphate. The control set were treated with a culture solution minus phosphate. Both sets were grown on through the summer period and leaf counts were made as a record of growth (Figure 16.2). This illustrates that the brief exposure to phosphate was effective, even outside the growing season. Phosphate storage could have a number of functions:

(a) To permit 'opportunistic' absorption of phosphate, especially in a system when pulses of phosphate availability may occur outside the growing season.
(b) To act as a sink for phosphate remobilised internally, particularly on senescence of organs. Root senescence may be of substantial importance, bearing in mind the relatively high biomass and its seasonal fluctuations.
(c) To act as a source for the production of seeds which have a sufficiently high phosphate content to provide for several months' growth on germination.

A WORKING HYPOTHESIS OF TERRESTRIAL OLIGOTROPHY

1. Defining vegetation types

Many treeless, shrub-dominated communities exist, and the question arises of their relationship to Australian heath. For example, mediterranean-type shrublands include the European types, *garigue* and *macchia*, the chaparral of California and its Chilean counterpart, and the *fynbosch* of South Africa. Outside the mediterranean climatic zone there are other shrublands. First, it must be remembered that some Australian heath is subtropical, tropical and montane. Then there are a wide variety of lowland and montane heathland types, a selection of which are described by Gimingham (1972). These are dominated by *Calluna* and other members of the Ericaceae. This vegetation type was first named 'heath' in common speech. As a type it extends from mineral soils to raised and blanket bogs, and is accompanied by the Cyperaceae as sclerophyllous graminoids. There appear to be transitional soil environments and vegetation types between 'heath', grassland or forest. A floristic link at least exists between the Cyperaceae of bogs and arctic tundra.

A bold step is to divide the mediterranean shrublands of Europe, California and Chile from all other heath types (Specht 1969). The small amount of environmental evidence that permits comparisons suggests that these formations are truly xerophytic and not especially adapted to extremely low phosphate substrates. They do, however, experience recurrent drought and rather frequent fire, and all include a number of symbiotic nitrogen-fixing types in the vegetation. There are many herbs in these ecosystems and a rather diverse group of shrubby angiosperms. This division would concentrate analysis on the following vegetation types:

(a) Australian heathlands as described;
(b) *fynbosch* of southern Africa, with many Proteaceae and also Restionaceae, Cyperaceae and Ericaceae (Adamson 1927);
(c) northern hemisphere *Calluna* heaths (Gimingham 1972);
(d) *Sphagnum* bogs in both hemispheres.

This extension of Specht's (1969) classification to include bogs is supported first by the presence of essentially heath vegetation on montane bogs in Australia. Secondly there is now abundant evidence from International Biological Programme studies that production in bog ecosystems is strongly phosphate limited.

2. The substrate

Using a 'unit–volume' notation to compare inorganic with organic substrates, a guideline, derived from many sources in the southern hemisphere and temperate Europe, seems to be in the order of 25–75 mg total P per litre of substrate. This ultra-oligotrophic condition may be derived by different pathways. In some cases extremely P-depauperate parent bedrock types are the fundamental cause; for example, the sandstones of southern Africa supporting *fynbosch*, or quartzites supporting lowland heath sites in eastern Ireland. In these examples it should be assumed that the weathering process has already occurred in the remote geological past. Weathering for long periods of time in more recent geological periods is a widespread cause, sometimes with the weathering being repeated, as in southern Australian and northern European deep sand deposits. A final process is that of paludification, or peat formation. One way of regarding massive peat deposits is to imagine that the organic matter serves as a diluent of any mineral nutrients in the soil. In raised bogs this may mean that the rooting zone is separated from the original mineral soil by some 10 m of peat. This has accumulated over a 10 000-year period. In blanket bog and hochmoor, a similar but less extreme condition exists. From a European perspective it is interesting to note the floristic links between lowland Australian heath and the montane *Sphagnum* bogs of southern Australia. This emphasises the separation between Australian heath and other 'mediterranean shrublands'.

3. Adaptations for oligotrophy should be sought

A series of adaptations contributing to an efficient phosphate economy, such as those described for Australian heath, will be common to all oligotrophic ecosystems. Many levels of study are obviously entailed, extending from ecosystem studies to phosphate ecophysiology.

A recently developed tool for the study of turnover is ^{31}P nuclear magnetic resonance (NMR) (Loughman and Ratcliffe 1984). This should greatly facilitate the study of both phosphate concentration and molecular state in tissue over the annual growth cycle. The need for ^{32}P studies and cumbersome biochemical separations may therefore be much reduced.

4. The taxa of oligotrophic environments

A limited and definable set of taxa would be involved. If an evolutionary trend is operating in the same way as that which generated the set of arid zone taxa (Table 3.1), then it should be possible to produce a select list of oligotrophic taxa. Table 16.2 represents a first approximation of such a list. The most striking feature of the list of orders conspicuously associated with oligotrophic sites is how few there are. It is also interesting to note the widespread distribution of Ericales and Cyperales. The presence of these two groups plus the carnivorous species is a fairly reliable marker for terrestrial oligotrophy.

At the present time a more coherent systematic pattern does not emerge. The very successful southern hemisphere groups, Proteales and Myrtales, may have had a wider distribution in Tertiary times, but in the absence of more palynological evidence the reasons for

Table 16.2: Select group of taxa commonly associated with oligotrophic substrates

Family	Representative genera	Notes
Cyperaceae	*Lepidosperma*	Australian heathland genus, which codominates
	Eriophorum	Bog and tundra genus which may dominate
		Both taxa are non-mycorrhizal
Restionaceae	*Hypolaena, Restio*	Southern hemisphere, in heath and *fynbosch*
Orchidaceae	Many	Minor but constant component worldwide. Not subjected to sufficient ecological analysis
Sarraceniaceae	*Sarracenia*	See Table 4.5. Terrestrial carnivorous groups are common to all the habitats noted. *Drosera* is especially widespread in heath and bog
Cephalotaceae	*Cephalotus*	
Byblidaceae	*Byblis*	
Droseraceae	*Drosera*	
Lentibulariaceae	*Pinguicula*	
Proteaceae	22 genera e.g. *Banksia*	Dominant in heaths and also found in bogs in Australia
	e.g. *Leucophyllum*	Dominant in *fynbosch*
Myrtaceae	38 genera, e.g. *Leptospermum myrsinoides*	Southern Australian heathland species
Epacridaceae	Many (28 genera) e.g. *Epacris*	Heath and bog shrubs in Australia
Ericaceae	e.g. *Erica* (many species)	Southern Africa and northern hemisphere
	e.g. *Calluna vulgaris*	Important dominant of heathlands and bogs in northern hemisphere
Fabaceae	24 genera e.g. *Phyllota*	Woody shrubs making a contribution to all heaths in Australia
	e.g. *Ulex*	Woody shrub in European heath (All are mycotrophic and possess nitrogen-fixing nodules.)

their present distributions are speculative.

The final speculative note to be made on this collection of orders is to observe that all are regarded as 'advanced' angiosperm types. Expressed crudely, they are the terminal branch ends of evolutionary trees constructed by systematists such as Hutchinson (1973), Cronquist (1981) and Takhtajan (1969). This is another ecotaxonomic avenue to explore, implying that plant evolution has also been conditioned by low fertility to produce these nutrient-efficient sclerophylls. The abundance of types in the southern hemisphere makes it a satisfactory place to begin this exploration. A biogeographical study of montane oligotrophic sites through the two Americas would enable a continuum to be constructed between the northern and southern floral elements.

17

Three aspects of the Alaskan Arctic tundra complex

The literature arising out of the IBP tundra biome studies will continue to provide stimulation for experiment and field study for many years to come. In particular, the two synthesis volumes associated with Alaskan Arctic tundra (Tieszen 1978; Brown, Miller, Tieszen and Bunnell, 1980) give us a variety of well-presented studies of soil–plant relationships in this 'simple' ecosystem.

The first tundra theme to be explored concerns the relationships between the patterned ground of the tundra landscape and the plant communities. This illustrates the interlocking of soil formation with biological process and climatic extremes. Nutrient cycling has been intensively studied in the sedge meadow vegetation types. This work is used to illustrate two further themes, namely (a) the control of nutrient availability as a key variable limiting plant production; and (b) interactions between the cyclic lemming population and the circulation of nutrients.

PATTERNED GROUND AND COMMUNITY DISTRIBUTION

The tundra biome vegetation comprises a series of eight vegetation formations distributed over a low-relief landscape of patterned ground. At Barrow, Alaska (71°N) the soil–parent material is fine-textured lacustrine or coastal alluvium, generally a silty clay or clay loam. The main soil-forming agencies have been the accumulation of organic matter and the complex pattern-forming processes arising from freeze–thaw cycles above permafrost. This has given rise to generally acid soils, with an organic-rich surface horizon and a horizon of clay or silt loam texture, passing to a perennially frozen

Figure 17.1: Climatic regime of the Barrow IBP site

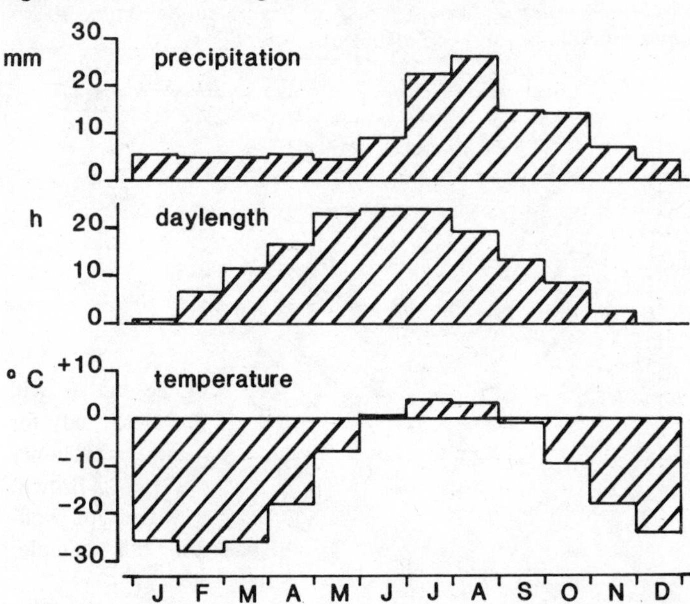

organic-rich horizon. The action of ice wedges penetrating many metres into the soil has generated a pattern of polygonal surface-relief features. These have a vertical amplitude of no more than 25–30 cm but have a profound effect on the flat and potentially meltwater-saturated landscape.

The climatic regime (Figure 17.1) of the Barrow site is such that energy balance rather abruptly changes from negative to positive in late May. Snow cover rapidly clears, exposing the shoots of the vegetation by mid-June. Some roots may remain close to freezing point for another 10 days. The maximum depth of thaw depends on the insulating properties of the above-ground biomass and the conductivity of the soil. Rapid shoot growth is possible, and soluble carbohydrate in plant tissues is abundant. Microclimatic effects on the topographic units are sufficiently great to cause differences in snow cover and depth and timing of soil thaw.

From Figure 17.2, the three major landscape units may be distinguished, the basins, rims and troughs of so-called 'low centred' polygon terrain. As the landscape develops, troughs broaden and deepen so that basins and rims are relatively high-relief features, giving rise to 'high-centred' polygon terrain.

All troughs may contain either ponds or streams. Older troughs

Figure 17.2: Patterned ground of the Arctic tundra in diagrammatic form. Above: a plan of the polygonal units. Below: a selection illustrating the effect of ice wedges

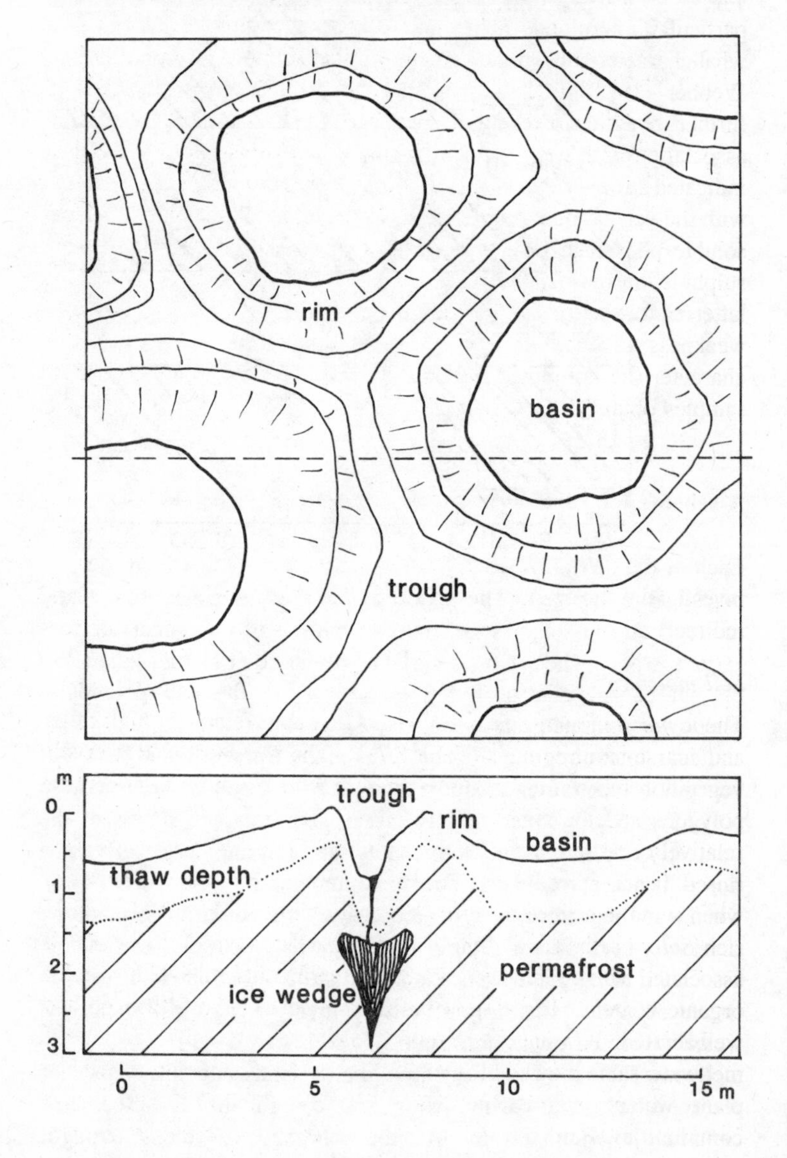

may broaden into drained but flat low meadow sites. A further subunit is associated with running water arising from late snowbanks and creek margins. The eight vegetation types are associated with particular microtopographic units. The key problem is to explain what characteristics of the soil environment control this distribution. Webber (1972) describes an approach which obviously leads to further hypothesis testing. He describes 16 substrate factors in association with each vegetation type. An ordination of these data indicated strongly that of the 16, three were most strongly associated with the vegetation pattern. These were soil moisure content, water-soluble phosphate and the extent to which the odour of hydrogen sulphide was associated with soil in the late growing season. This latter characteristic is a crude measure of redox status. The major weakness of the study is the semi-quantitative nature of this characteristic combined with the rather low number of phosphate samples obtained.

A scheme for further hypothesis testing

Each of the three prime correlates in pattern forming needs further investigation to establish if the relationship is truly causal or indirect.

Soil moisture

The low soil moisture sites are associated with relatively high relief and coarse-textured soils (Table 17.1). The two most characteristic vegetation types are *Luzula* heath and *Salix* heath on high-centred polygons and the rims of low-centred polygons respectively. The relatively scarce *Luzula* heath, found also on some coarse-textured raised beaches, may experience authentic water stress, possibly when wind exposure occurs while soils are still frozen. The forb-rich *Salix* heath has a slightly longer period of snow cover and is associated with a rather wide range of sandy–silty soils with variable organic content. These may be interpreted as relatively deep and well aerated. To what extent roots avoid water stress by reaching the meltwater table may be of interest. Most of the fieldwork on soil and plant water potential has been carried out on other tundra communities. Before water stress is seen to be a barrier to colonisation of 'dry' sites by other suites of species, this should be remedied.

Table 17.1: Soil characteristics associated with low soil moisture

	Unit	June snow depth (cm)	Aug max. thaw depth (cm)	Organic component (cm)	Surface organic thick-ness (cm)	Sand/silt ratio %
Luzula heath	High-centred polygon	0	37	24	4.4	13
Salix heath	Rim of low-centred polygon	4	54	12	3.7	34

Soil aeration

The index of aeration based on the odour of hydrogen sulphide is probably better than first thoughts would indicate. The human nose is a good detector of H_2S at low concentrations, and H_2S is generated at a clearly defined redox potential (see Chapter 9). When first detected, the percentage of soil volume generating H_2S may be quite low, with the rate of generation controlled by temperature. As odour is perceived to increase, this implies a larger source size or increasing soil volume. However, if H_2S is perceivable, then dissolved oxygen in soil water is zero and oxygen for root growth is supplied via the well developed aerenchymas of the dominant graminoid species. Presence of H_2S may therefore indicate that conditions are not suitable for species with insufficient plasticity of root development to permit aerenchyma formation. Mycorrhizal roots would also function poorly under these conditions.

A better picture of soil aeration would be obtained by determining the pattern of oxygen diffusion rate and redox potential with time and soil depth. Influence of potentially toxic ferrous and manganous ions also bears investigation.

It is of interest that the graminoids have highest biomass where soil moisture, hydrogen sulphide and soluble phosphate are all high. This may indicate the direct connection between low redox potential and the release of phosphate ions from iron phosphates.

Soluble phosphate

This was the only soil fertility variable studied, and although selected with some knowledge of the powerful effects of phosphate limitation on growth, it begs the question as to the influence of other nutrients. Availability of nitrogen is a particular case. A more complex fertility index may be more appropriate. Phosphate supply may well be mediated by different mechanisms in different vegetation types. There may also be an interplay between nitrogen, phosphate and cationic nutrient elements.

For example, it is suggested above that phosphate release is conditioned by low redox potential in graminoid-dominated sites. Absorption and assimilation of phosphate occur along the length of non-mycorrhizal but aerenchymatous roots. The situation in forb- and shrub-dominated sites may be very different, with freeze–thaw cycles in early and late season giving soluble phosphate pulses. In between a large volume of aerated soil with active mineralisation is explored by a root–mycorrhizal complex. Phosphate uptake may occur without high soluble phosphate being detected in the soil. Nitrogen availability may be equally complex, and rates of nitrification, ammonification and nitrogen fixation are worth investigation. It may be that they are directly linked to the phosphate situation.

NUTRIENT LIMITATION TO GROWTH OF SEDGE MEADOW VEGETATION

Sedge-meadow types form some 75% of the Barrow tundra vegetation and as such are the best studied from the nutrient utilisation standpoint. A series of vegetation types, most containing *Carex aquatilis* (Cyperaceae) with varying quantities of two *Eriophorum* species (Cyperaceae) and two Gramineae, *Dupontia fisheri* and *Poa arctica*. Small changes in topography are expressed by a mosaic of these vegetation types. This mosaic constitutes the unit of study to be considered.

Evidence for limitation to growth in sedge meadow arising from nitrogen and phosphorus supply comes from a number of sources:

(a) Highest production occurs in the sites with highest soluble phosphate concentration.
(b) Applications of nutrients give a marked growth response.
(c) Nutrient concentrations of tissues are quite low.
(d) Phosphate absorption continues throughout the season. It is noted that 40% of seasonal phosphate absorption occurs after net downward translocation has commenced. Absorption is only terminated when soil freezes in September. This indicates that phosphate storage in roots occurs, sufficient to permit nutrient loss by grazing and providing a supply for early season growth.

Figure 17.3: Phosphate budget (mg P per m^2 per year) for tundra sedge meadow, redrawn from Chapin, Miller, Billings and Coyne (1980)

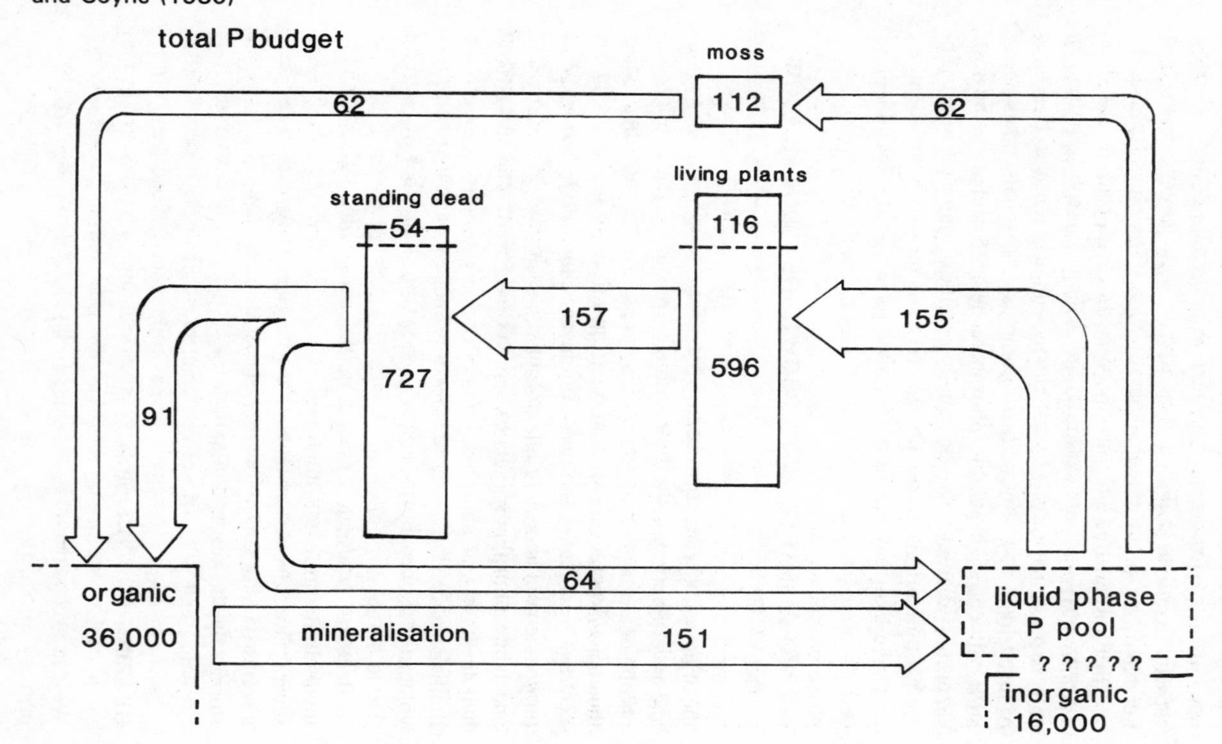

The soils associated with sedge-meadow vegetation have a reasonably high total phosphate-P content, namely 250 mg P per litre of soil (0–10 cm). The largest soil phosphate pool is organic phosphate, an index of limitations to mineralisation. Activities of mineralisers are limited by temperature and soil aeration. Their capacity to release phosphate may also be curtailed by high C:P ratios, i.e. microbes may be responsible for phosphate immobilisation.

The phosphorus budget for one unit of sedge meadow, the *Carex aquatilis–Oncophorus wahlenbergii–Dupontia fisheri*-dominated unit, is presented in Figure 17.3. This diagram combines the work of many tundra biome scientists and it is redrawn from the compilation of Chapin, Miller, Billings and Coyne (1980). In this simplified version, phosphate inputs and losses are omitted. They are negligible, being less than 1 mg per square metre per year.

The two solid-phase pools, organic and inorganic, are very large, being some 64 times the size of the living plant pool. The 'liquid-phase phosphate pool' is not quantified in size, except that a little over 200 mg phosphate per square metre per year passes through it to supply the needs of vascular plants and mosses. This quantity, passing through a 'standing dead' phase, is eventually transferred to the organic phosphate pool. Mineralisation studies on fungi, bacteria and invertebrates indicate that a quantity in the order of 200 mg phosphate per square metre per year is released. This release would thus appear to account for the demand of the plants.

What is uncertain is the linkage between the liquid-phase phosphate pool and inorganic phosphate, even though exchangeable and labile phosphate fractions can be measured. The most likely hypothesis is that phosphate mineralisation rate is the key growth-limiting factor in this unit of study. The constraints are (a) the total volume of thawed soil; (b) the temperature regime; (c) poor aeration due to soil texture and waterlogging; and (d) low pH.

Because of such factors as changing soil moisture and invertebrate grazing, microbial population 'crashes' may occur several times during the season. Thus mineralisation may proceed in a series of pulses. This further emphasises the need for phosphate storage by the vascular plants.

GRAZING AND THE SEDGE-MEADOW NUTRIENT CYCLE

We can accept from the foregoing material (a) that sedge-meadow primary production is nutrient limited; and (b) that there are complex interactions between the mineralisation of nutrients, soil conditions and climate. The next question to be asked is 'How do tundra grazing animals affect this system?' The Barrow system is somewhat simplified in that the brown lemming is the principal vertebrate grazer as other species have been virtually eliminated by hunting. Unlike many grazed systems where grazing pressure remains reasonably uniform, lemming populations fluctuate from a baseline density of less than one animal per hectare to a peak of 150–200 every three to six years. The increase is due to a slight improvement in survival of adult females and their young. Catastrophic mortality explains the radical decline. Massive disruption of the vegetation occurs during a peak, with clipping of graminoids and disturbance of moss and lichen carpets. It was also fairly clear that mortality had both starvation and predation as components. Of the many possibilities that might explain the population fluctuation, one hypothesis has been carefully examined over the last 20 years, the nutrient recovery hypothesis (Pitelka 1964; Schultz 1964).

The nutrient recovery hypothesis

The principal ideas of the nutrient recovery hypothesis are outlined in Figure 17.4. Its key purpose was to arouse interest in the whole of the system and not merely in aspects of breeding biology and population dynamics, predation or energetics.

How does the hypothesis rest after some 20 years of study (Batzli, White, Maclean, Pitelka and Collier, 1980)? First, the building phase of a population increase does result in high mineralisation, especially of phosphate. Soluble phosphate released from the overwinter faeces is estimated to be as high as 90 mg P m^{-2} (compare with Figure 17.3). The reduction in biomass by grazing, at peak density, is about 50%, except where rhizomes are consumed when it may be as high as 90%. Thinning of plant cover does increase thaw depth, but by a small amount. However, most of the root biomass is still close to the soil surface despite increased thaw depth. In fact, other evidence indicates that when thaw depth is increased by other means, nutrient availability and plant growth increase.

Some other explanation must be sought for the observed decline in phosphate concentration. This may lie in the depletion of reserve phosphate, drained by replenishing grazed shoot tissue. Furthermore, when increases in plant tissue phosphate were observed, this was not followed by increases in population as predicted by the

Figure 17.4: The 'nutrient recovery hypothesis' which seeks to explain the linkage between lemming population cycles and vegetation. See text for a comment on its validity

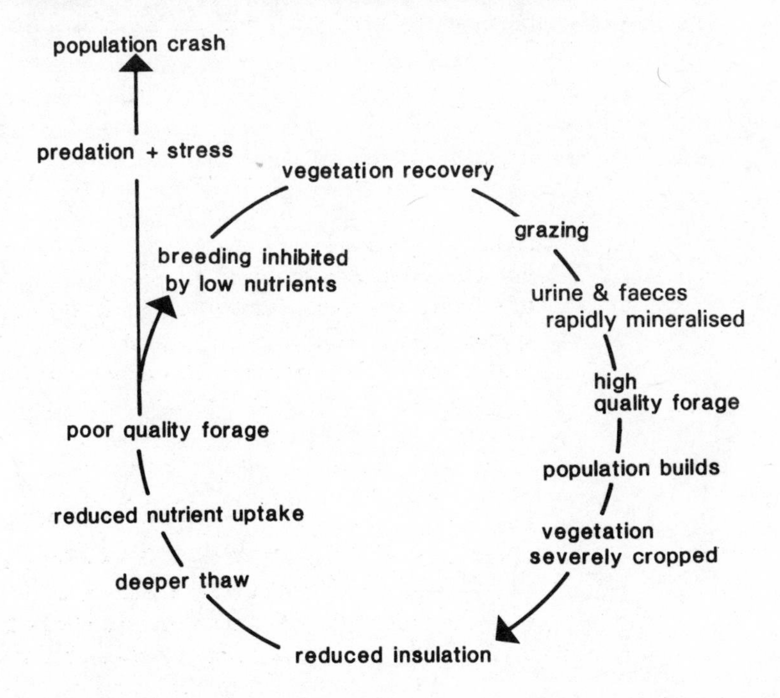

nutrient recovery hypothesis.

Another aspect of lemming activity is the capacity of the animals to redistribute nutrients through the landscape units. Winter nesting habitat is the polygon trough where deep snow improves their survival opportunities. Nutrients are removed from basins by foraging and are deposited in troughs adjacent to winter nests as urine and faeces. Summer nests are located on rims, which achieve an intermediate nutrient status after troughs. The nutrient-depleted basins provide a less attractive forage.

This view is largely hypothetical and needs testing and development. Predators may also have a function in nutrient redistribution, in particular the winter predation of weasels encouraging nutrient concentration in troughs and avian predators, which may be responsible for high phosphate concentrations on high-centred polygons.

It is now quite clear, in general, that predation by herbivores is definitely connected with increased rates of nutrient cycling. A more exciting speculative end point is that the interaction between vegetation, nutrients and predators reinforces the formation of tundra patterned-ground.

18

Saltmarshes and the coastal zone

Vegetated lands regularly inundated by tidal waters have both a biological fascination and considerable economic importance. One aspect of biological interest comes from considering a set of terrestrial plant taxa collectively termed 'halophytes', which have accommodated to the marine environment. These range from the grasses and rushes, many herbs and subshrubs, to the diverse woody taxa of the mangroves of the world. This must be one of the best examples of convergent evolution in the plant kingdom.

A second biological feature is the link that these communities form between terrestrial and coastal ecosystems. On the one hand halophyte communities are productive and can contribute detrital particulates to tidal water. Conversely, sediments continually accreting from tidal waters are the parent material for most intertidal soils. The nature of this exchange requires to be understood in qualitative and quantitative terms.

An understanding of ecosystem dynamics is an important requirement from a resource management viewpoint. The resources in question are:

- intertidal and coastal fisheries
- assimilative capacity for wastes, especially sewage
- grazing lands
- timber and firewood production
- grass-fibre production

This is a management problem common to the developed and developing worlds. It is made more urgent by the world-wide pressure on estuarine and coastal locations for urbanisation and industrialisation.

A further feature of interest is the role these communities play in coastal geomorphology. The shape of many low-lying coastlines is a result of interactions between sediment, tide and intertidal vegetation.

From a large set of possibilities, three aspects are presented:

(a) an analysis of a vegetation zonation in temperate saltmarsh;
(b) a review of halophyte ecophysiology;
(c) a nitrogen budget for an estuarine ecosystem.

AN INTERPRETATION OF SALTMARSH VEGETATION ZONATION

Most accounts of the vegetation of European saltmarshes refer to a zonation generally corresponding to degree of tidal cover, e.g. Chapman (1960), Ranwell (1972) and Beeftink (1977). The following account, based on a saltmarsh in Dublin Bay, is an attempt to construct a hypothesis that explains the zonation.

Substrate

Wherever stable conditions of relative sea level exist, saltmarsh formation occurs under well defined geomorphological situations. This generally means a shore (a) sheltered from destructive waves or currents; (b) supplied with suspended sediment in the tidal water; and (c) given frequent periods in which sediment deposition can occur. In Dublin Bay these conditions are fulfilled on the lee shore of a 5.5-km-long sandbar island, North Bull Island (Jeffrey, Healy, Holland and Moore 1977).

In the section of the system displayed (Figure 18.1a) some 200 m of permanently vegetated saltmarsh has formed in a little more than 100 years. Other sections in the literature are incomplete, more complex or of uncertain age.

The sediment has a number of distinct components:

(a) Fine sand: this material predominates in the subtidal and lower intertidal areas and as beach and dune. However, in the saltmarsh substrate, sand grains comprise a small fraction of the total, usually less than 1%. The exceptions are either sandy lamellae, interpreted as marking storm events, or the presence

Figure 18.1a: Vegetation zonation, topography and substrate on a temperate saltmarsh, North Bull Island, Dublin Bay. Section of saltmarsh showing topography, thickness of silt-rich soil and vegetation cover. Composition of vegetation is simply presence in contiguous 1 m^2 quadrats. Data from McNamee (1976)

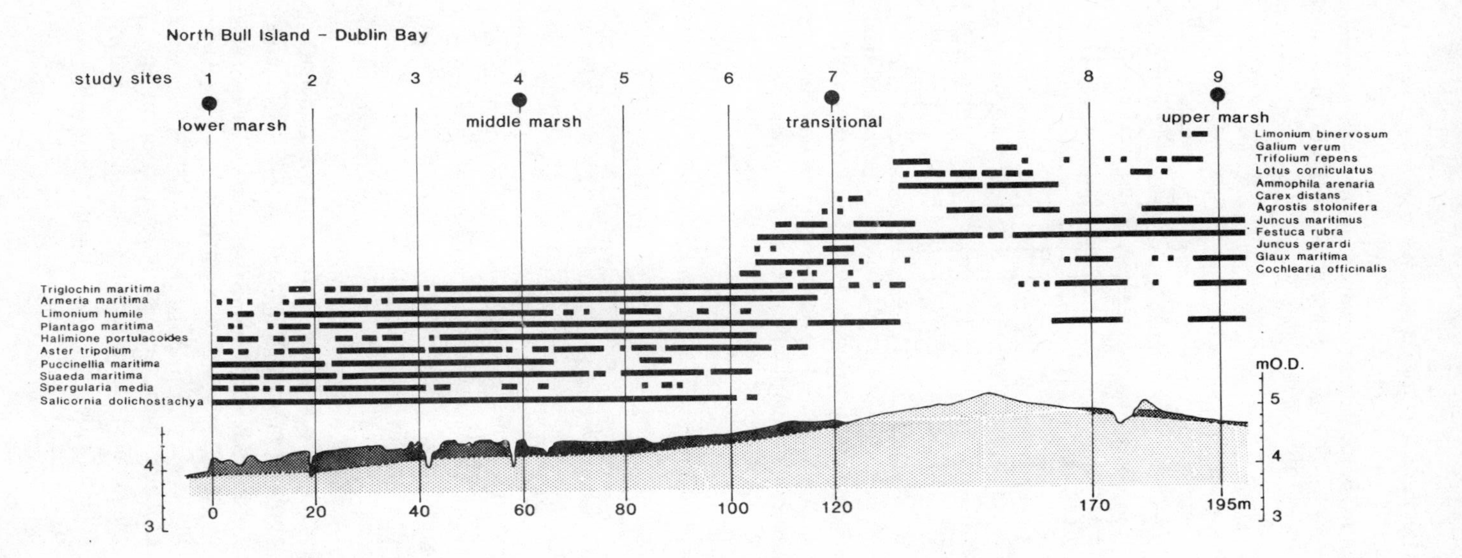

Figure 18.1b: Biomass and nitrogen content of vegetation samples collected from the study sites in Figure 18.1a. Shoot is shown at the upper end of each bar

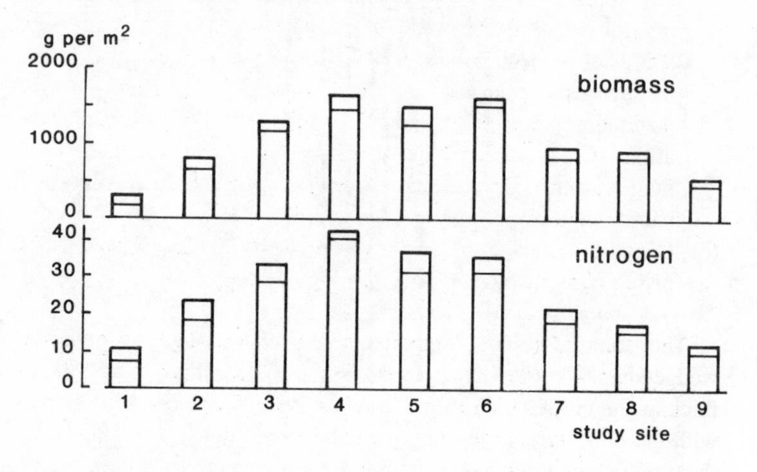

Figure 18.1c: Contribution of species on a percentage biomass basis to the vegetation of the three most distinctive parts of the saltmarsh

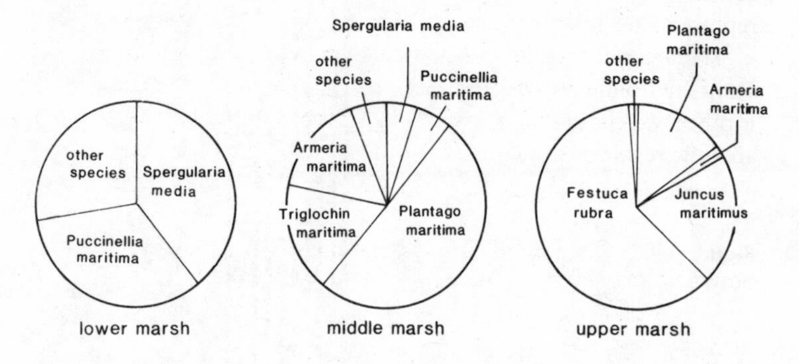

in the upper saltmarsh from blown sand dune. Sand is not normally borne suspended in slow-moving spring tide waters.

(b) Silt is the predominant mineral fraction, and consists mainly of silica carried in suspension from rivers flowing into Dublin Bay.

(c) Calcium carbonate is a component arising from finely comminuted shells. Although classified as sand by sieve separation, the flake-like shell fragments tend to behave as silt in sedimentation. A substantial proportion of the inorganic phosphate is associated with these aragonite surfaces.

(d) Detrital organic matter. This is a very variable feature, and not easy to quantify exactly because of the difficulties of separating organic input from estuarine primary production. The main source of detritus is the freshwater inputs into the bay, amplified by sewage discharges. Organic detritus is codeposited with silt throughout estuarine systems. In industrial or urbanised situations it may carry ligand-bound metals or include emulsified hydrocarbons. Organic input from sediment is also an important source of nitrogen.

(e) Organic matter from primary production, as in all soils, is a prime consequence of soil development.

The four sediment components are codeposited to give the wedge-shaped section depicted in Figure 18.1a. This is some 30 cm thick at the eroded cliff at the seaward side. To landward it blends with the dune sand at the limit of exceptional tides. The reason for the diminishing thickness is simply the reduction in tidal inundations which carry silt and detritus into the system. However, the highest spring tide of a series sweeps a drift line of floating debris and detritus to the highest parts of the system, where it probably represents an input of mineralisable nutrients, especially nitrogen.

The unusual process of soil formation generates a strongly zoned set of environments whenever it occurs. These are reflected by some form of vegetation zonation, e.g. Figure 18.1a. Both observations are now regarded as commonplace ideas but there is still uncertainty as to which environmental factors have the most direct effect. Three

Figure 18.2: Diagram illustrating the possibilities for interaction between soil salinity, soil aeration and soil fertility

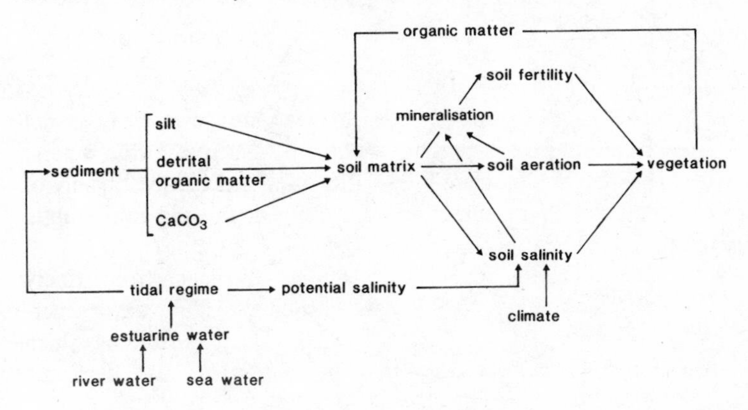

environmental characteristics appear to predominate: (a) soil salinity, (b) soil aeration, and (c) soil fertility, especially soil nitrogen status. All these factors are complex in themselves, and there is ample opportunity for interactions (Figure 18.2)

Salinity

The ionic composition of sea water is shown in Figure 18.3, which emphasises its origin as soluble products of crustal weathering. As a plant growth medium it is deficient in the major nutrient elements nitrogen and phosphorus, and has a high ionic strength due mainly to Na^+ and Cl^- ions. The osmotic potential of coastal sea water, approximately −2.4 MPa, combined with the toxic effects to metabolism of Na^+ and Cl^-, serve as a barrier to the invasion of the intertidal zone by normal terrestrial vegetation. Even the vegetation of non-saline wetlands is excluded. Saltmarsh vegetation is conferred with the ‘privilege’ of avoiding competition from the general regional flora while incurring an ‘obligation’ to evolve systems for coping with a novel environment for vascular plants.

Before the characteristics of halophytes are discussed, the range of environmental salinity under consideration should be noted. If −2.4 MPa is a norm for the osmotic potential of sea water, it is clear that the interstitial soil water may vary on either side of this value. A range of psychrometrically determined soil osmotic potentials from a saltmarsh in Dublin Bay illustrates this (Figure 18.4)

Figure 18.3: Ionic composition of sea water

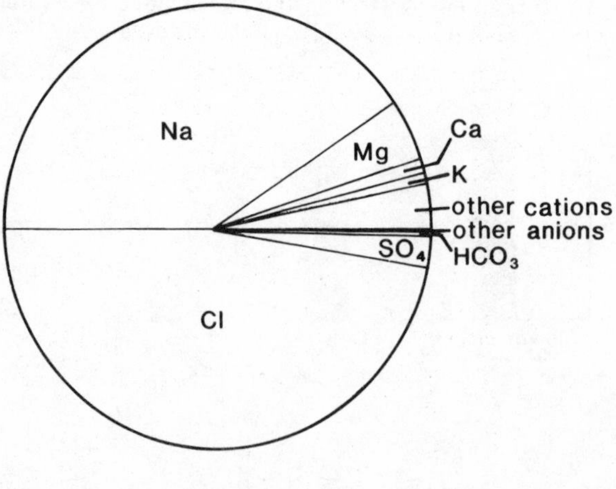

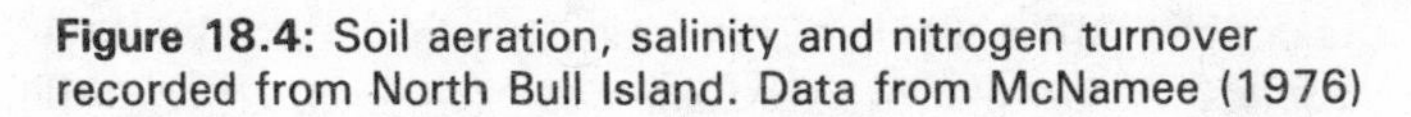

Figure 18.4: Soil aeration, salinity and nitrogen turnover recorded from North Bull Island. Data from McNamee (1976)

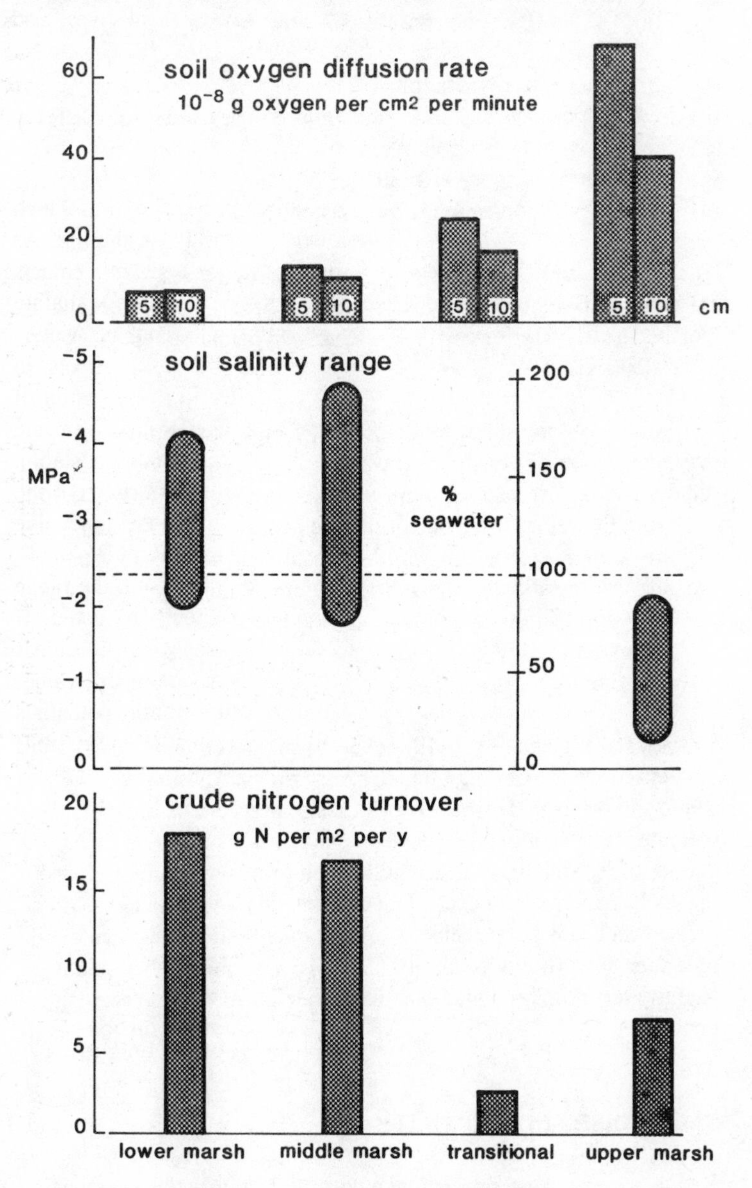

(McNamee and Jeffrey 1977). The lowest soil salinity recorded, the bulwark against invasion by non-halophytes, lies in the region of −0.2 to −0.3 MPa. Low salinity appears to be a result of:

(a) infrequent tidal inundation;
(b) low soil permeability and hence failure of sea water to penetrate the soil column from above;
(c) wash out of salts by rainfall;
(d) influence of non-saline groundwater.

At the other end of the scale, salinities of twice sea water (−4.8 MPa) have been recorded from time to time on this site and in Norfolk (UK) (Jefferies 1977). Such hypersaline episodes are generated by:

(a) relatively long duration of cover by high spring tides;
(b) penetration of soil by tide water;
(c) long exposure periods, which, in the case of Dublin Bay, result from the rather low tidal amplitude occurring in May and June;
(d) the influence of all the climatic factors promoting evaporation and transpiration, i.e. maximum hours of sunshine, highest soil and vegetation temperatures, coastal breezes and relatively low relative humidities;
(e) low rainfall.

Salinities as high as −4.8 MPa revert to sea water salinities with the next exposure to a spring tide. Sea water is thus a diluent of hypersaline soil water under these conditions. Observations in Dublin Bay indicate that a doubling of sea water salinity is a rare event, but some degree of hypersalinity is a regular feature of the summer environment. It may never occur under some climatic regimes. At the present time is is not possible to predict quantitatively the generation of either low or high salinity. This is a problem that might be usefully taken up by a mathematical modeller of eco-meteorological conditions.

SALTMARSH SOIL FERTILITY

By many standards the intertidal environment ought to be reasonably fertile. Sea water contains many essential ions in abundance, i.e. K^+, Ca^{2+}, Mg^{2+}, SO_4^{2-}. Sediment is buffered towards alkalinity (pH 8.5) and usually contains trace elements (iron, manganese, zinc). Organic matter contains a certain amount of the total nitrogen.

The principal nutrient of importance in terms of limiting fertility appears to be nitrogen. There are several reasons for this view:

(a) total nitrogen is not exceptionally high;
(b) accumulation of organic matter suggests limited mineralisation;
(c) nitrogen fixation is at a low level (Jones 1974);
(d) annual species respond to added nitrogen under greenhouse conditions (Pigott 1969);
(e) low nitrate reductase activity is a feature of saltmarsh plants, except those low in the marsh (Stewart, Lee and Orebamjo 1973);
(f) denitrification appears to be a major process in estuaries in general (Valiela and Teal 1979) and saltmarshes in particular (Nedwell 1982).

On the other hand, addition of nitrogen to high saltmarsh does not give a marked growth response (Jefferies and Perkins 1977; Sheehy-Skeffington 1983). This lack of responsiveness may be due to the relatively high salinities encountered, the metabolic storage capacity of vegetation referred to previously, or the removal of soluble nitrogen by tide water.

Differences in the nitrogen status of saltmarsh zones may be determined from three areas of evidence:

(a) The nitrate reductase evidence of Stewart *et al.* (1973). The argument is that since NR is inducible, high enzyme concentrations in tissue are an index of high rate of supply. Lower zones in the saltmarshes investigated by these authors exhibited high enzyme activities.
(b) Measurements of nitrogen availability by incubation methods confirm that availability is relatively high in the lowest zone.
(c) Estimates of nitrogen turnover indicate that the highest turnover occurs low in the marsh. The turnover measurement is simply the difference between highest and lowest zonal biomass multiplied by the tissue nitrogen concentration for the appropriate time. The pattern of estimated turnover (Figure 18.4) is not reflected in the tissue concentrations or even in the standing crop of nitrogen. The turnover value is a minimum quantity of nitrogen required to sustain the observed production. It neglects any nitrogen lost during the period of

biomass accumulation by grazing or losses to litter.

A major missing link in any understanding of nitrogen supply is lack of knowledge about the soil organisms and microbes contributing to mineralisation. Casual direct observation indicates that saltmarsh soils are dominated by bacteria and nematodes, with fungi and other faunal types not well represented.

The aeration factor in saltmarshes

In all soils the concentration of oxygen at a given point is the result of oxygen supply rate on one hand versus the rate of oxygen consumption on the other. Fine-textured silty soils plus organic detritus, when saturated with saline water, generate conditions leading to: (a) slow gas diffusion through fine water-filled pores; (b) microbial consumption of organic matter. Hence organic-rich inter-tidal muds are frequently anaerobic with such low redox potentials that free S^- ions and H_2S are apparent. A major process in the global sulphur cycle is the reduction of SO_4^{2-} in sea water by estuarine heterotrophs under these conditions, which ultimately reacts with ferrous iron to give black FeS of low solubility.

The only process leading to increased aeration of blackened muds is the burrowing of tolerant animals such as the annelid worm *Peloscolex*. Colonisation by plants is probably limited to types such as *Spartina* (Armstrong 1982) or mangroves. Penetration of anaerobic mud by *Spartina* roots increases oxygen flow into the substrate via their aerenchyma to a limited extent. This may be insufficient to protect the roots when high sulphide production occurs (Goodman and Williams 1961). This seems to be the reason for *Spartina* 'dieback disease' in southern England.

Higher in the saltmarsh sequence aeration improves. Diffusion channels are provided by living and dead roots and the cracking of soils on drying. As the organic matter content increases, the soil becomes free draining, and in marshes with much *Festuca rubra* anaerobic conditions are seldom encountered. The improvement in soil aeration is thus almost entirely a function of the activity of plants.

Adverse conditions for plant growth consequent on the decreased redox potential may be deduced from Table 9.1. First Mn^{2+} (manganous ions) then Fe^{2+} (ferrous ions) and finally S^{2-} (sulph-ide ions) would be released as the redox potential falls from above

Table 18.1: A hypothetical model of the causes of saltmarsh zonation in terms of salinity, soil aeration and nitrogen supply

Zone	Low	Middle	High
Salinity	Sea water	< Sea water to 2 × sea water	< Sea water
Nitrogen	High turnover	Medium turnover	Low turnover
Aeration	Low redox potential and long periods of poor aeration		Aeration good

+400 mV to −150 mV. All three ions have been stated to have toxic properties in the saltmarsh context (Havill, Ingold and Pearson 1985; Rozema, Luppes and Broekman 1985; Singer and Havill 1985). The outcome of these experimental studies is that upper marsh species are more sensitive to all three ions. Middle and lower marsh species demonstrate variable sensitivity. Thus the spread of species from upper to middle marsh may be inhibited by low aeration. It must be remembered, however, that aeration and redox-potential-linked ion release may be seasonal and spasmodic characteristics.

A model that combines the three factors is set out in Table 18.1.

THE HALOPHYTE STRATEGY

Analysing the salinity environment from a halophyte viewpoint there are three basic specifications:

(1) the need to develop a water potential gradient from soil, through all the cellular compartments of the plant, to the air;
(2) a requirement for manipulating an ionic balance normally hostile to plant metabolism, i.e. high ratios of Na^+ and Cl^- to other essential ions;
(3) capacity for accommodating to short-term fluctuations in soil salinity.

The water potential gradient problem has been subject to a great deal of study, especially in the range zero to −3.0 MPa. Halophytes clearly achieve an appropriate scaling of the ψ gradient with leaf tissue ψ more negative than the soil, a ψ_p in the region of + 0.5 MPa (Stewart and Ahmad 1983) and negative xylem potential in wood of mangrove species (Scholander, Hammel, Bradstreet and Hemmingsen

1965). Leaf cell sap osmotic potentials are largely accounted for by NaCl, but there may be a substantial gap to account for. Solutes that may fill this gap are suggested by a long list of rather unusual metabolites known to accumulate in halophytes (see Table 18.2)

Relationships between cytoplasm and vacuole

Studies of the osmotic potential of expressed cell sap with respect to substrate salinity are a relatively crude measure of adjustment. The relationships between cytoplasm and vacuole are also of great importance. The vacuole confers turgor on tissues and is a site of metabolite storage, and metabolic properties depend on the integrity of the cytoplasm. Sampling of the osmotic potential of the whole tissue is biased towards the 90% of solution-filled cell volume comprising the vacuole. Nevertheless, the cytoplasm must also be in water potential equilibrium with both cell wall and vacuole, as swelling or desiccation of cytoplasm is not observed.

It is widely accepted (a) that the cytoplasm has a low ionic content with respect to the vacuole: a ratio of 1 : 5 is quoted by Stewart and Ahmad (1983); (b) that enzyme systems of halophytes are no less sensitive to NaCl than non-halophytes. This implies that the halophyte maintains a low Na cytoplasm. If ions do not contribute much to the low water potential of halophyte cytoplasm, what type of molecule does, especially without inhibiting metabolism?

Compatible metabolites

A picture is building up of osmotically active substances, which do not inhibit metabolism, that are regularly detected in the tissues of halophytes (Table 18.2 and Figure 18.5). These may be termed compatible metabolites. Temperate saltmarsh species were the first types to be investigated (e.g. Jefferies 1980, Gorham, Hughes and Wyn Jones 1980), but now the cataloguing activity has been extended to the subtropical and tropical mangroves (Popp 1984a,b; Popp, Lahrer and Weigel 1984).

A classification of herbaceous and woody halophytes has now been erected on the basis of dominant organic solutes, the latest version being presented by Popp *et al.* (1984) (see Table 18.2, which also includes data from Briens and Lahrer (1982) and Gorham *et al.* 1980).

It is anticipated that the process of biochemical and chemotax-

Figure 18.5: Structure of some compatible metabolites contributing to osmotic potential in halophytes

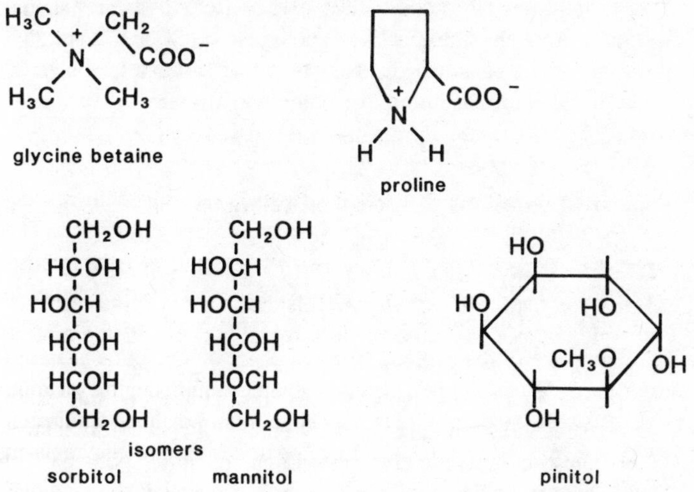

glycine betaine

proline

isomers

sorbitol

mannitol

pinitol

Table 18.2: Compatible metabolite compounds associated with halophytes

(1) Methylated quaternary ammonium accumulators, e.g. glycine betaine: Mangrove spp., *Avicenna marina, A. eucalyptifolia, Hibiscus tiliaceus, Heritiera littoralis; Atriplex hastata, Salicornia europaea, Suaeda macrocarpa*

(2) Proline accumulators: *Triglochin maritima*; two mangroves, *Xylocarpus gronatum* and *X. thekongensis*

(3) Proline and methylated quaternary ammonium accumulators: *Agropyron pungens; Limonium vulgare; Spartina alterniflora;* one mangrove sp., *Acanthus ilicifolius.*

(4) Proline and miscellaneous solute accumulators: *Triglochin maritima* (reducing sugars)

(5) Sucrose and/or polyol accumulators: *Puccinellia maritima, Plantago maritima* (sorbitol), *Aster tripolium, Juncus maritimus, Phragmites communis*; five mangrove spp. (mannitol), *Aegiceras corniculatum, Luminitzera littorea, L. racemosa, Sonneratia alba, Scyphiphora hydrophylacea*

(6) Cyclitol accumulators: seven mangrove spp. (pinitol); all Rhizophoraceae; *Honkenya peploides; Spergularia media*

(7) Other compounds — unidentified carbohydrates: *Agropyron pungens, Aster tripolium, Festuca rubra, Puccinellia maritima*

onomic inventory will continue, possibly with new classes of organic solutes added. The classification in Table 18.2 will certainly be further modified. The compatible-metabolite hypothesis raises a collection of new physiological and ecological questions. A number of these questions relate to the adjustment of salinity in the field, which will conclude discussion of halophyte strategy.

Physiological questions of ecological relevance

(1) *Does the osmotic metabolite fulfil the function of buffering the osmotic potential of cytoplasm?* It is clear that in some cases the organic compounds cited are at too high a concentration in the tissue as a whole to be confined to cytoplasm, e.g. in the case of pinitol in certain mangrove tissues. However, studies of compartmentalisation are not well enough advanced to answer this question completely. Providing osmotic potential of cytoplasm is sufficiently low, it does not matter if the vacuolar osmotic potential is derived from a mixture of organic solutes and ions.

(2) *Which mechanisms initiate and control the concentration of compatible metabolites?* This is a real problem because of the fluctuations in substrate osmotic potential encountered in the field. A realistic working range for substrate would be −0.1 to −5.0 MPa. A preliminary approach to this problem could use a mixture of a non-ionic osmotic substrate, e.g. polyethylene glycol, with sea-water ions.

(3) *What are the metabolic sources and sinks for compatible metabolites?* It is easy to conceive of a reversible nonsoluble-to-soluble carbohydrate transformation generating the low-molecular-weight solutes and subsequently reducing their concentration. The high underground biomass of some species, e.g. *Plantago, Armeria, Triglochin*, may provide a reservoir of stored metabolites. However, not all types have such an asset, e.g. the grasses.

Ecological questions

(1) *Does the system of osmotic buffering permit exact adjustment to changing substrate salinity?* Most of the literature on

biochemistry and chemotaxonomy is only broadly related to ecological details. No substrate osmotic potentials are quoted in the literature on the compatible metabolites cited in Table 18.2. An interesting physiological study was carried out with cell cultures of the saltmarsh grass *Distichlis spicata* (Daines and Gould 1985). This demonstrates that Na^+ and proline permit an accommodation to salinising treatment within 12 hours.

(2) *Are there implications for the nitrogen economy of coastal communities?* Nitrogen in proline and glycine betaine may contribute up to 20% of the nitrogen in a leaf. There seem to be at least three ways in which a plant may conserve as scarce a resource as nitrogen in a saltmarsh: (a) resort to more carbohydrates rather than the nitrogenous compounds; (b) have a ceiling for all nitrogenous compounds used as osmotic buffers (Briens and Lahrer 1982); (c) recycle compounds via storage tissue. All these possibilities require further study.

(3) *Do the compatible metabolites confer any other ecological benefit?* Some physiological information suggests that glycine betaine, proline, sorbitol and mannitol all confer protection to enzymes and membranes against heat damage (Popp *et al.* 1984). Whether there will be a movement towards a unified theory that brings together responses to certain extreme environments is an interesting question. Tolerance to high temperatures and sub-zero temperatures, desiccation and salinity could all conceivably have a common physiological basis.

So far there is evidence that proline, glycine betaine, sorbitol and mannitol increase the heat stability of enzymes extracted from sand-dune species (Smirnoff and Stewart 1985). There is a great deal to be done before anything like a unified theory can be thought of as widely plausible. The presence of compatible metabolites may also play a role in discouraging herbivores.

Throughout this section some effort has been made to avoid the word ‘stress’, which appears so frequently in the literature on halophytes. For the intertidal or salt-desert halophytes, soil solutions with a substantial osmotic, rather than matric, potential are the norm. Halophytes have acclimated to this unusual environment in a specialised way and should not be viewed as being under ‘stress’. ‘Constraint’ is perhaps a better word, as salinity certainly does offer a barrier to growth. ‘Stress’ is a term best reserved for an environmental effect that is totally unusual, e.g. the effects on cereals of irrigation with saline water. In this situation we are observing an ecò-pathological stress effect, with a minimum of adaptive physiological flexibility.

Adjustment to fluctuating salinity in the field

The mere persistence of vegetation in the face of seasonal salinity fluctuations is a token that adjustment occurs. In the literature already cited there is also evidence of salt and compatible metabolites fluctuating in step with experimental alteration in environmental salinity. These tracking systems have not been examined in the field or studied under conditions leading to hyper-salinity. The key questions are:

(1) What are the sources and sinks for both ions and organic metabolites in tracking salinity changes?
(2) What are the environmental cues for initiating change?

The answers to these questions must be speculative and comprise information from disconnected fragments of the literature.

Salt glands, present in some but not all types, may permit dumping of NaCl, especially when tracking a period of declining salinity. *Armeria maritima* and various *Limonium* species (Luttge 1971) can pump saline solutions to the epidermal surface.

Storage tissues may form another part of the answer. High root biomass (Figure 18.1b; McNamee and Jeffrey 1977) may be a characteristic of the regions experiencing highest salinity. In the case of *Plantago maritima* experiencing an increasing salinity, leaves senesce and are rapidly replaced by new leaves with a modified osmotic regime. We would speculate that the root provides a depot for translocated materials and a source for the fabric of the new tissue plus the sorbitol present as a compatible metabolite. It is also of interest that leaf morphology alters during this transformation (McNamee and Jeffrey 1977).

As to the triggering or osmoperception mechanism, it can be seen that changes are not initiated if osmotic potential changes are applied using polethylene glycol or if saltmarsh plants are droughted, even in the absence of NaCl. They tend to die. This may simply be that salt is an essential part of the adjustment process, but at the same time osmotic metabolite synthesis needs to be stimulated.

Figure 18.6: Sketch map of Great Sippewissett Marsh, Cape Cod, Massachusetts. This indicates the nearly 50 ha unit of study. *Spartina alterniflora* marsh occupies 44% of the area with a fringe of *Spartina patens* (18%). Creeks running between the *Spartina* beds occupy 35%, and scattered salt pans and algal mats some 3%. To the north and east the site is bounded by glacial drift upland and a railway embankment. To the west is a dune-covered sand spit. Groundwater and other freshwater inputs are represented by arrows. Tidal exchange occurs with Buzzards Bay, the southern embayment of Cape Cod. The amplitude of spring tides is approximately 1.5 m

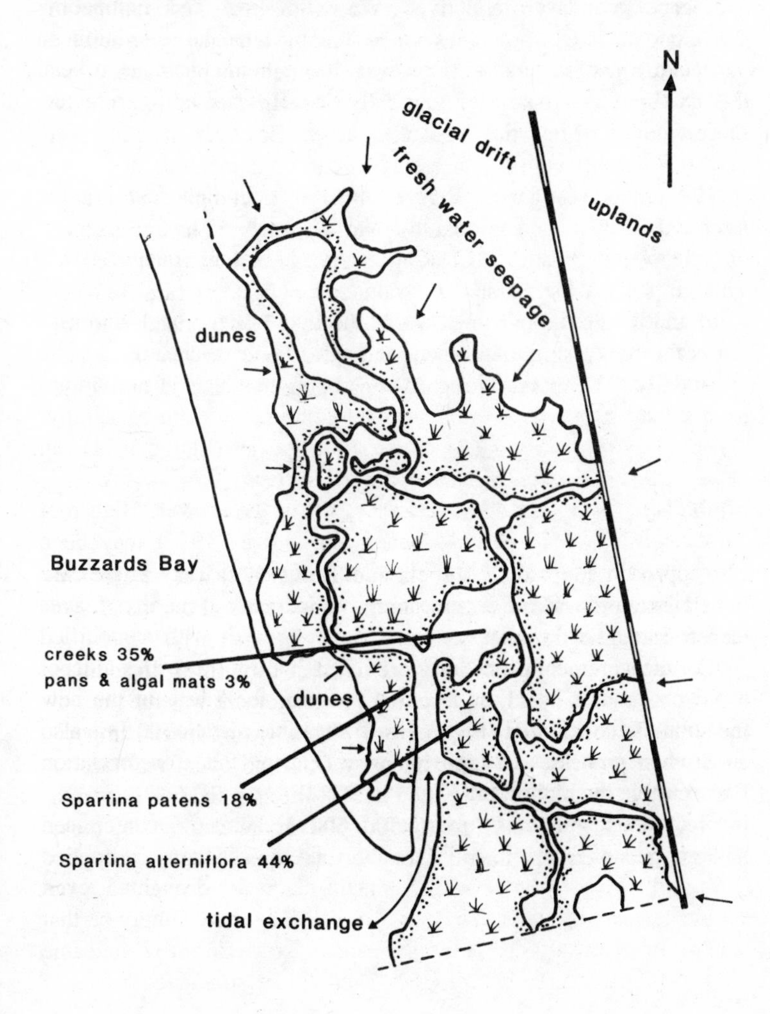

NITROGEN BUDGET OF A SALTMARSH ECOSYSTEM

Background

The importance of this study (Valiela and Teal 1979) is that it represents the first attempt to undertake a mass balance of nitrogen in a vegetated coastal ecosystem. Previous coastal studies have indicated the limitation by nitrogen of coastal primary production and hence secondary productivity. We will use these estimations to gain a perspective of processes concerning the nitrogen cycle and the particular role of vegetation. However, it should not be assumed that this example is a basis for generalisation. In fact it may not be representative of intertidal coastal systems. However, it is the best yardstick against which future studies can be compared.

The unit investigated (Figure 18.6) is a complex of creek, intertidal mudflat, and saltmarsh covering nearly 50 ha and located on the western shore of Cape Cod, Massachusetts, USA. Topographically the marsh is surrounded by glacial moraine to landward and by a dune barrier towards the sea. The connection to the sea is through a single entrance, and tides flood the marsh system twice daily. The marsh is regarded as being mature and in a long-term steady state.

Methods

The approach adopted by Valiela and Teal (1979) was to examine in detail the processes operating in the subsystems of the marsh and then to integrate the results.

The measurement procedures are listed in Table 18.3. In addition to the procedures listed, an important measurement was the area of the subunits considered (Figure 18.6). This was of special importance when processes of high intensity occurred over small areas. For example the highest absolute rates of nitrogen fixation occurred in algal mats and pans (up to 500–600 ng N cm^{-2} h^{-1}); these habitats occupied less than 3% of the total area.

Table 18.3: Input and output processes and their measurement

Process	Measurement	Notes
Groundwater flow	Spring water samples and calculation of total flow by measurement of salinity reduction, NO_3, NO_2 NH_4, dissolved organic nitrogen (DON) particulate nitrogen (PN)	Some agricultural and domestic sewage assumed. Total input 6120 kg N g^{-1}
Precipitation	Rain-gauge collections. Analysis as above	7.9 kh N ha^{-1} y^{-1} is within range quoted for temperate-zone sites
Nitrogen fixation	Separate investigation of subsystems using acetylene reduction technique	Highest rates were in small areas of algal mats contributing 9%. The most important system as a whole was rhizosphere-associated bacteria (80%). Free-living bacteria contributed 11%
Tidal water exchange	Detailed hydrographic measurement of tidal flows across entrance channel accompanied by water sampling and analysis as precipitation and groundwater	DON is the major component but there is evidence that some 96% is refractory and not utilised. Seasonal rhythms are important for other components
Denitrification	Measurement by gas chromatography of nitrogen released into helium-flushed containers	Strong seasonal pattern associates with temperature. Creek bottoms accounted for 50% of denitrification. High fixation in algal mats was matched by high denitrification. Nitrate is not exported in spring and summer when denitrification is most active
Losses to sediment	Defined as losses below 15 cm depth, especially beneath vegetation. A sedimentation rate of 0.15 cm g^{-1} was assumed	This amounts to about 4 mg N m^{-2} y^{-1} for vegetated areas
Minor gains and losses	NH_3 volatilisation estimated from literature. Input from roosting gulls' defecation estimated by census and analysis. Shellfish harvests estimated from records	Total is less than 0.1% of inputs or outputs

Table 18.4: Overall annual nitrogen budget for 48.3 ha marsh system (kg N y^{-1}). (Coefficients of variation about 20% of the mean values presented.)

Process	Form of nitrogen								
	NO_3	NO_2	NH_4	PN	DON	N_2	Other	Total	%
Input									
Precipitation	110	0.4	70	15	190	—	—	385.4	1.07
Groundwater flow	2 920	30	460	—	2 710	—	—	6 120	17.00
N_2 fixation	—	—	—	—	—	3 280	—	3 280	9.11
Tidewater	390	150	2 620	6 740	16 300	—	—	26 200	72.80
Bird faeces	—	—	—	—	—	—	9	9	0.02
Total input	3 420	180.4	3 150	6 755	19 200	3 280	9	35 994.4	
%	9.50	0.50	8.75	18.77	53.34	9.11	0.03	—	100.0
Output									
Tidewater	1 210	170	3 540	8 200	18 500	—	—	31 620	79.29
Denitrification	—	—	—	—	—	6 940	—	6 940	17.40
Sedimentation	—	—	—	—	—	—	1 295	1 295	3.25
Ammonia volatilisation	—	—	17	—	—	—	—	17	0.04
Shellfish harvest	—	—	—	—	—	—	9	9	0.02
Total output	1 210	170	3 557	8 200	18 500	6 940	1 304	39 881	
%	3.03	0.43	8.92	20.56	46.39	17.40	3.27		100.0
Balance	+2 210	−10.4	−390	−1 445	+800	−3 360	−1 295	−2 495.4	

Results

The table of results (Table 18.5) illustrates all the input and output data. Interpretation of it is hampered by the large input and output of refractory DON. A better perspective is gained by representing this by a net input value of 800 kg N y^{-1}, the difference between input and output values (Tables 18.4 and 18.5; see also Figure 18.7).

The marsh is an importer of oxidised nitrogen and DON and a net exporter of particulate nitrogen, some ammonium and a large amount of dinitrogen gas. The higher plants absorb and reduce inorganic nitrogen to produce particulates and nitrogen which is mineralised to ammonium. About 40% of the net annual above-ground production of the vegetation is exported as particulates. The value

Table 18.5: A summary table to indicate the main transactions

	Input	Output	Net	Net as % of input
NO_3 and NO_2-N	3 600	1 380	+2 220	+62
NH_4-N	3 150	3 557	−407	−13
DON	800	See text	+800	+100
PN	6 755	8 200	−1 445	−21
N_2	3 280	6 940	−3 660	−112
Totals	17 585	20 077	−2 492	−14

Figure 18.7: Schematic diagram of main nitrogen transactions for Great Sippewissett Marsh

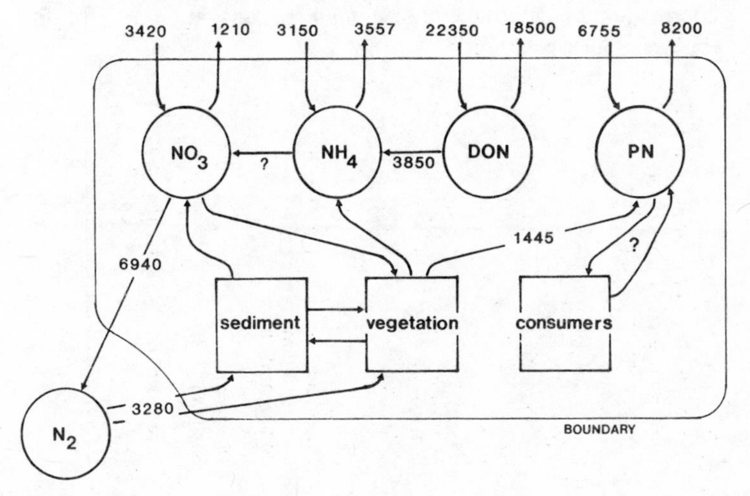

of the marsh is thus assessable in terms of resource exported to coastal ecosystems, namely PN, and its capacity for denitrification. The authors speculate on the capacity of the system to cope with an increased load of polluted groundwater.

In terms of the biology of the nitrogen cycle, this example shows first that multiple inputs and outputs can occur. Their multiplicity is probably a special feature of this case. Secondly, the separation of processes in space and time is of interest, notably of nitrogen fixation and denitrification. Thirdly, we should note the differing intensity of processes, e.g. high-intensity nitrogen fixation in algal mats, and medium-intensity fixation in the rhizosphere systems of *Spartina*, the latter process being the more important in the whole system. Lastly, it should be noted that the two grass genera, *Spartina* and *Distichlis* are C-4 species, having a greater efficiency of growth per unit nitrogen. This is suggested as a deterrent to potential herbivore predators. This aspect of the system was not investigated, but it seems probable that senescence, death and microbial decomposition are the pathway to particulate organic nitrogen (PN) rather than predation and faecal pellets.

Comparisons between this system and many other intertidal situations will eventually be made, and conclusions will be drawn as to the generality of results. A key variable may be the nature of the tidal flushing regime. In a more truly estuarine situation, with a substantial freshwater through flow, a greater flux of particulate organic nitrogen to the ocean may result. Destruction of intertidal vegetation by development, it is argued, will deplete protein resources accessible to the coastal food web. We need more information to determine the strength of this argument for the conservationist position.

19

Calcareous and serpentine soils and their vegetation

INTRODUCTION

In these cases a very conspicuous ecological pattern is imposed by substrate on vegetation. This pattern virtually transcends all other environmental factors, and thus in some ways resembles the effect of phosphate depletion on Australian vegetation (see Chapter 16). However, in both these cases the environmental factors concerned are multiple, generally adverse to optimum plant growth and interactive, and differ in intensity from site to site. The ecological analysis of either case is by no means complete. Historically, interest focused first on the peculiarities of the vegetation, with environmental interpretation following. Now that some sense of proportion is emerging with respect to the patterns of the soil environment, it seems reasonable to start with them. This can then be followed by an explanation of the response by the vegetation. In both cases the environment imposed by the soil leads to selection and adaptation by the flora. At the same time the environment provides a refuge from competition by acting as a barrier to less adaptive species. This is by now a familiar pattern.

The analysis of calcareous situations in Europe is in a reasonably mature state, with the gaps in the state of knowledge well defined. In contrast, the serpentine case is still wide open, despite a profuse and well reviewed literature. It seems a healthy state of affairs to conclude with a modest, but speculative, proposal towards solving a problem.

CALCAREOUS SOILS AND THEIR VEGETATION

Calcium carbonate

The formation and dissolution of solid-phase calcium carbonate is a recurrent feature of the geochemistry of the Earth's crust. Calcium is a plentiful element (Table 7.1), most of its salts are slightly soluble in water, and Ca^{2+} is the most abundant ion in soil solution. Many natural waters, including some soil solutions, the ocean and some fresh waters, are nearly saturated solutions of calcium salts. The biological carbon cycle interacts with calcium-containing solutions either by generation of carbon dioxide (respiration) or by removing carbon dioxide (photosynthesis).

Formation

Precipitation of calcium carbonate, as calcite, occurs when carbon dioxide is photosynthetically extracted from a calcium bicarbonate solution. Calcite, the most common trigonal crystal form of $CaCO_3$, may be found impregnating the cell walls of marine and freshwater macro-algae, e.g. the coralline red algae and the Characeae. Calcareous cell walls of the planktonic algae, the Foraminifera, have in particular given rise to the 'chalk' deposits of the Cretaceous period. Most important of all is the extracellular precipitation that occurs in productive marine environments. This has given rise to massive oceanic calcite deposits, often remarkably pure, which lead in the course of geological time to limestones. This may be expressed by the equilibrium equation:

$$2CO_2 + 2H_2O \rightleftharpoons 2H_2CO_3 \rightarrow 2HCO_3 + Ca^{2+} \rightarrow CaCO_3 + H_2O$$

$$2H_2CO_3 \downarrow 2H^+ \qquad 2HCO_3 \searrow \text{Photosynthetic } CO_2 \text{ absorption}$$

A second biological process contributing to $CaCO_3$ formation is the very highly controlled formation of calcified structures by many groups of organisms, especially corals, echinoderms and molluscs. Here the crystal form of $CaCO_3$ is often aragonite, an orthorhombic crystal which is slightly more soluble and reactive in soil. Fossilisation tends to lead to conversion to calcite. The composition of limestones may depend first on the sources of $CaCO_3$, secondly on the circumstances of sedimentation, and thirdly on the subsequent geological history. These will all have a bearing on the properties of the rooting substrate. Calcium carbonate deposition may also occur by simple evaporation of calcium-bicarbonate-containing solutions. This leads to the formation of rock types such as travertine and tufa and also to $CaCO_3$ in arid-zone soils.

A range of calcareous soil-forming materials exist, which are reviewed in Table 19.1. It is certainly worth analysing for $CaCO_3$ in soils with a pH greater than 6.0.

Dissolution of calcium carbonate in soils

Dissolution is important in two ecological respects: (a) it explains how decalcification can occur and why essentially acidic rooting horizons may overlay a calcareous base; and (b) in the short term it provides an explanation of the buffering by $CaCO_3$ of the soil liquid phase.

Ion absorption

Soil CO_2 partial pressure — Cation exchange complex

$$H_2O + CO_2 \rightleftharpoons H_2CO_3 \rightleftharpoons H^+ + HCO_3^- \rightleftharpoons CO_3^{2+} + Ca^{2+} \rightleftharpoons CaCO_3 \text{ solid}$$

When high carbon dioxide concentrations are generated, hydrogen ion concentrations and $CaCO_3$ dissolution rates are high. Thus surface decalcification of a low carbonate soil is associated with continued decomposition of organic matter. When much $CaCO_3$ is present, the pH is stabilised, with a pH maximum of 8.3. This value corresponds to the general atmospheric CO_2 concentration of about 300 vpm. Also stabilised are Ca^{2+} and HCO_3^- concentrations in the liquid phase. However, the equilibrium may shift if soil CO_2 partial pressure increases.

Surface absorptive reactions

The surface of a $CaCO_3$ crystal in the soil, with a dynamic equilibrium with soil solution as depicted above, may serve as a sink for other ions. In agriculture phosphate added as fertiliser is immobilised by the formation of apatite crystals associated with the carbonate surface. Zinc and manganese ions may also be immobilised on the carbonate surfaces.

The bulk of soil inorganic phosphate may be assumed to be in the form of minerals of the apatite series. However, it must be remembered that the stability/solubility of these minerals is dependent, at a given pH, on the ionic composition of the crystal lattice.

Table 19.1: Substrates giving rise to calcareous soils

(A) *Massive limestone and chalk deposits*
Throughout the geological column from Cambrian to recent. A wide range of structures, lithologies and degrees of metamorphism exist. Because of internal weathering and drainage, 'karst' or 'cockpit' landforms may develop by solution and collapse. Geological variants important to soil formation are:

(1) Dolomitisation in which the deposit is enriched with $MgCO_3$
(2) Hydrothermal enrichment with elements, usually as vein-filling minerals such as CaF_2, fluorspar; PbS, galena; various zinc minerals
(3) Addition of silica, either cosedimented, giving horizontal so-called 'chert' horizons, or as veins
(4) Coprecipitation or addition on metamorphosis of apatite minerals, i.e. $Ca_{10}(PO_4.CO_3)_6(OH.F)_2$
(5) Cosedimentation of sand or silt or clay
(6) Incorporation of pyrite, FeS_2, or hydrated iron oxides
(7 Metamorphism by heat giving rise to massively recrystalline hard marbles or friable 'sugar limestones'
(8) Folding and faulting giving rise to specific landscape types and more local features controlled by joints, e.g. 'clints' and 'grykes' of limestone pavements

(B) *Transported calcareous deposits*
Derived from the above, but may overlie non-calcareous bedrock. Types include boulder clays with a wide range of unsorted particles; loesses, which are wind-sorted and transported periglacial materials. These obviously need not be calcareous and sometimes form a contrasting cover to limestone bedrock. Calcareous gravels and finer alluvial sediments are transported by rivers across bedrock boundaries

(C) *Coastal soils*
The soils of coastal dunes and saltmarshes usually contain aragonite from wave-comminuted 'recent' shell fragments. Some coastal deposits may consist entirely of eroded limestone, coral, shell or algal carbonate fragments. Decalcification of silica sands with less than 10% $CaCO_3$ frequently generates a vegetation pattern

(D) *Freshwater marshes and fens*
Fine-grained and shelly calcareous deposits accompany photosynthetic activity in base-rich shallow lakes, termed 'marl lakes'. Marl, a mixture of calcareous mud with organic matter, is often the substrate for a succession of vegetation types in the temperate zone, the 'fen' formation

(E) *Arid zone soils*
Calcium carbonate, arising from evaporation of soil solutions, may be accompanied by gypsum and possibly other salts. Osmotic potential may have some influence on vegetation, and if sodium salts are present pH may rise beyond pH 8.3

Table 19.2: Calcareous soils as rooting media

Negative features of calcareous soils include:

(a) Absolute quantities of major nutrients may be low, especially phosphate, potassium and nitrogen
(b) Nutrient availability, where directly pH controlled, is less than optimum, especially phosphate, iron, copper, zinc and manganese
(c) Cation exchange capacity is low, because of general absence of layer silicates. Soil organic matter is the main reservoir of CEC
(d) Microbiological activity is suboptimal, leading to reduced nutrient turnover and mycorrhizal activity
(e) Coarse soil textures and shallow development may contribute periodic water stress due to poor water storage capacity

On the positive side it must be recognised that:

(f) Available Ca^{2+} is always high, giving the plant superoptimal conditions for membrane function and cell development
(g) Soil aluminium is low, both absolutely and in availability terms
(h) Available nitrogen is in the form of nitrate
(i) When lead, copper and zinc are present at elevated concentrations, availability is low
(j) Most soils are freely draining and seldom waterlogged

The presence of iron or aluminium phosphates should not be ruled out, even in the most calcareous soils.

If calcareous soils are taken as a generality, the potential properties of the rooting zone may be summarised (Table 19.2). The soil properties outlined are intended to indicate that calcareous soils are extreme habitats. As most extreme habitats, they have encouraged the development of specialised floras; and have prevented colonisation by less specialised vegetation. Furthermore, attention is drawn to a habitat type, the antithesis of calcareous soils, characterised by low pH and its consequences. Here an equally specialised flora exists.

Knowledge of these floras and of their close correlation with calcareous or acidic substrates has been one focal point of ecological study in Europe. The reason for this is that patterns of geological substrates are extremely clear cut and relatively small in scale. When describing vegetation, geological discontinuities are frequently encountered and the effects on vegetation discernible within a few metres. On a continental scale there is a close relationship between calcareous substrates and the presence or absence of

altitudinal zone	soil type: calcicolous	soil type: mesophilous	soil type: silicicolous
alpine — upper		swards with Carex curvula	
grasslands	Sesleria grasslands		Festuca & Nardus grasslands
marshes	Carex davalliana & Kobresia simpliciuscula		Carex fusca & Eriophorum spp.
snowbeds	Arabis caerulea & Salix reticulata		Salix herbacea
rockfaces	Androsace helvetica & Saxifraga diapensioides		Androsace vandelii & Saxifraga florulenta
scree	Thlaspi rotundifolia		Oxyria digyna
2000m — subalpine	Pinus uncinata	Larix europea	Picea abies
1200–1500m — montane	Pinus sylvestris		Fagus sylvatica
550–800m — hill	Quercus pubescens	Carpinus; Ostrya	Quercus robur; Alnus incana
upper mediterranean	Quercus ilex		
lower mediterranean	Pinus halipensis	Quercus suber; Ceratonia siliqua	Populus alba

Figure 19.1: Synoptic diagram of the vegetation of the European Alps which focuses on calcareous and silicaceous substrates. Simplified after Ozenda (1983)

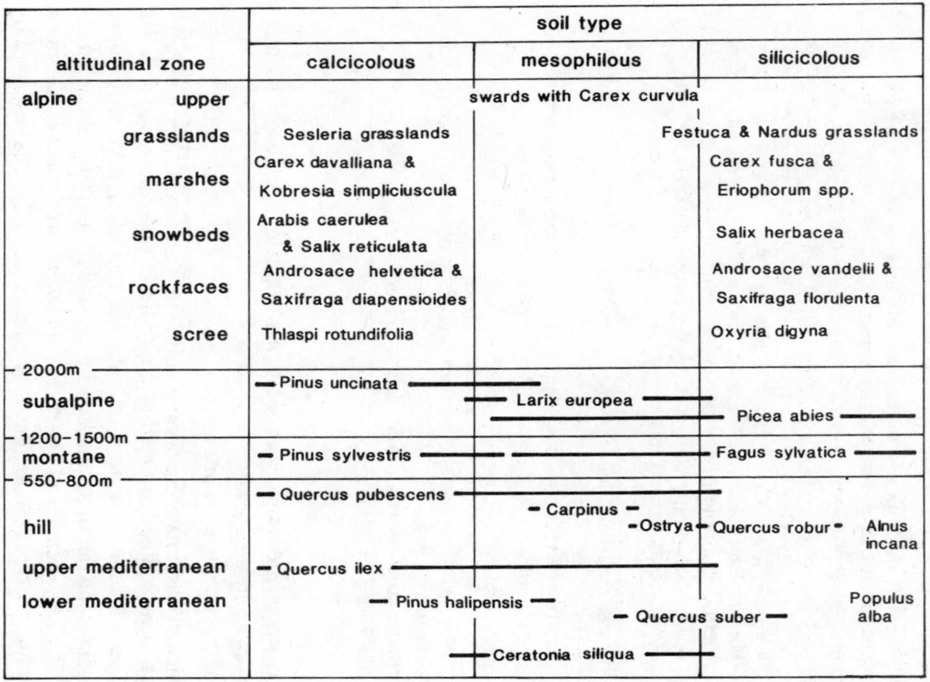

particular species. The term 'calcicole' or 'calciphile' (lime loving) has been applied in the English literature to species with a distribution correlated positively with calcareous substrates. They are not good terms because of the outmoded idea that these species have an overwhelming requirement for a resource, rather than a tolerance of adverse factors. Species with a positive correlation for soils of low pH and low base status are more satisfactorily termed 'calcifuges', 'fleeing' from lime. The alternative term 'silicicole' has been coined.

A good example of the power of contrasting soil type is seen in the synoptic diagram of the vegetation of the European Alps (Ozenda 1983). A simplified scheme is presented in Figure 19.1. The author points out that the word 'calcicolous' implies also 'xerophilous' and 'silicicolous' implies 'hygrophilous'.

Another function of this diagram is to remind us that 'calcicolous' and 'silicicolous' vegetation types are not mere alternatives but the extreme ends of a continuum. Between them are a series of vegetation types which comprise the generality of plant cover, Ozenda's 'mesophiles'. Critical experiments frequently compare types from both ends of the continuum, when one end against the middle might prove more enlightening.

CALCICOLES AND CALCIUM CARBONATE CONTAINING SOILS

Calcium carbonate containing soils are widespread and embrace a wide range of possible environmental associations:

(1) Frequently dry and well aerated but sometimes moist and waterlogged.
(2) Low-fertility thin soils are often encountered but deep fertile soils also occur.
(3) Geological anomalies occasionally give rise to high toxic-element concentrations.
(4) All climatic and altitudinal situations may be encountered.

What are the constant environmental features that apply over this range?

(1) Calcium ion supply is sustained.
(2) Bicarbonate ion is always present.

(3) Iron availability is always minimal.
(4) Aluminium is never present in toxic concentrations.

Two questions must be answered:

(a) Do these characteristics confine calcicoles to their substrate?
(b) Do these characteristics form a barrier to colonisation by other elements of the regional flora?

Calcium in supply and demand

Analysis of tissues of calcicoles indicates high Ca^{2+} concentrations, probably interpreted as 'luxury' absorption. Any tests of high Ca^{2+} demand must be carried out under carefully controlled culture experiments. In the rather few species investigated, mainly seedling forbs, calcicole types give a growth response to calcium additions. Moreover, this response is absent in non-calcicole species that exhibit depressed growth.

Calcicoles may both demand high Ca^{2+} and have a tolerance of very high concentrations. For non-calcicoles, once the rather low and easily satisfied essential calcium demands are met, Ca^{2+} may well behave in a toxic fashion. A summary of the available physiological literature (Kinzel 1983) suggests that cytoplasmic Ca^{2+} must be maintained at a low level by ATP-driven Ca^{2+} efflux pumps. If the pumping capacity is overwhelmed, then a metabolically destructive Ca^{2+} concentration results. Choice of species may be important in exploring this hypothetical view. To compare calcicoles with types ecologically confined to extreme acid and nutrient-poor soils may not yield the most ecologically meaningful results. For example, there are many possible contrasting species within the large alpine genera *Campanula, Primula, Saxifraga* and *Dianthus*. The latter two taxa contain species that eliminate or immobilise calcium (Rorison and Robinson 1984). Calcium toxicity thus has a case to answer in restricting mesophiles.

Is bicarbonate of importance to calcicoles?

There is no evidence that suggests that bicarbonate has strong direct effects on calcicoles. An indirect influence is the possibility of an interaction with iron availability. However, there is some evidence that bicarbonate ion restricts root elongation in non-calcicoles (Lee and Woolhouse 1969).

Are calcicoles affected by low iron availability?

The solubility of $Fe(OH)_3$ decreases by about 1000 times for each increment of pH between 4 and 8. Calcicoles growing in the range pH 7–8.3 seem to be very 'iron efficient' (Brown 1978), in that the element may be obtained and utilised under these conditions.

A series of screening experiments are described by Grime and Hodgson (1969) which show categorically that calcicoles are 'iron efficient' to the extent of not being seriously affected by 'lime-induced chlorosis'. The symptoms are a marked yellowing to bleaching of leaves, which may be transient or permanent. Amelioration is obtained by treatment with application of chelated iron and resembles the iron deficiencies met in agriculture and horticulture. The syndrome entails more than inability to absorb iron at high pH. Utilisation of absorbed iron appears to be inhibited, with bicarbonate playing a role in the inhibition process. It is suggested that in the bicarbonate-rich environment of the rooting zone 'iron inactivator' compounds form in the presence of phosphate. Whether a phosphate effect can operate in ecological, in contrast to horticultural, situations is open to doubt (Kinzel 1983). At present the physiology of lime-induced chlorosis in non-calcicoles is not understood.

There are three different, but not mutually exclusive, hypotheses that can account for the iron efficiency of calcicoles:

(a) pH at the absorbing (root or mycorrhizal) surface is lowered. Very local pH changes could occur as H^+ ions are extruded counter to cation absorption or associated with respiratory CO_2 release.

(b) Reducing conditions at the root surface, giving rise to readily absorbed Fe^{2+}, could promote uptake. An interesting variant of this phenomenon is the possibility that the centres of soil aggregates become anaerobic, while the bulk soil is well aerated. Exploration of aggregates by absorbing surfaces could be an effective tactic, even if the fine roots and mycelial hyphae had a short life.

(c) Chelated iron is relatively easily absorbed, and natural chelating agents such as the dicarboxylic citrate ion are known to be found in root exudates.

An interesting study of the iron-efficient *Helianthus annuus* (Romheld and Marschner 1979) showed that a positive response occurs to the onset of iron deficiency. Reducing capacity of the root surface increases and pH is lowered simultaneously. Uptake and translocation of iron, from ^{59}Fe-labelled $Fe(OH)_3$, increased by 100-fold. Morphological changes were also noted, including root-hair proliferation. These changes are interpreted as a fine tuning of the assimilation of iron which can recur as the iron status of the root alters with growth.

Future work may reveal more widespread examples of a similar set of responses and probably of different responses adapted to particular subenvironments. Whatever the full ecophysiological explanation, iron efficiency does seem to be a key adaptation necessary for colonisation of all but the wettest calcareous soils.

Are calcicoles aluminium sensitive?

In mineral soils Al^{3+} ions become massively available at pH 4.5. Some calcicoles are known to be sensitive to less than 5 ppm Al^{3+}. Symptoms include a set of leaf symptoms resembling phosphate deficiency, and roots which are stunted and brown in colour. The toxic effect arises from the intense cross-linking associated with the trivalent ion. Affected systems include cell division and ion absorption (Clarkson 1984).

It is not difficult to screen species for aluminium sensitivity with a seed germination assay on filter paper. Sensitive species react to Al^{3+} solutions by a stunting and browning of the newly emerged radicle. A convenient characteristic is the Al^{3+} concentration needed to cause 50% inhibition of root growth in one-week-old seedlings. This was used by Grime and Hodgson (1969) to test a range of grassland plants (Figure 19.2). A simple plot of the aluminium sensitivity against soil pH of the seed collecting site indicates that not only calcicoles are inhibited by low aluminium concentration. The data clearly show that only species occupying soils of pH < 4.5 possess tolerance to aluminium. Both calcicoles and mesophiles occupying the middle range of soil pH, i.e. 4.5–7.0, are Al^{3+} sensitive.

An interesting case is that of *Schoenus nigricans* (Cyperaceae), an aluminium-sensitive calcicole confined to base-rich sites in England. In Ireland, however, the plant is frequently found as a component of blanket-bog vegetation on acid peat. The paradox is

Figure 19.2: Aluminium sensitivity of seedlings with respect to pH of seed collecting site. Plotted from data of Grime and Hodgson (1969)

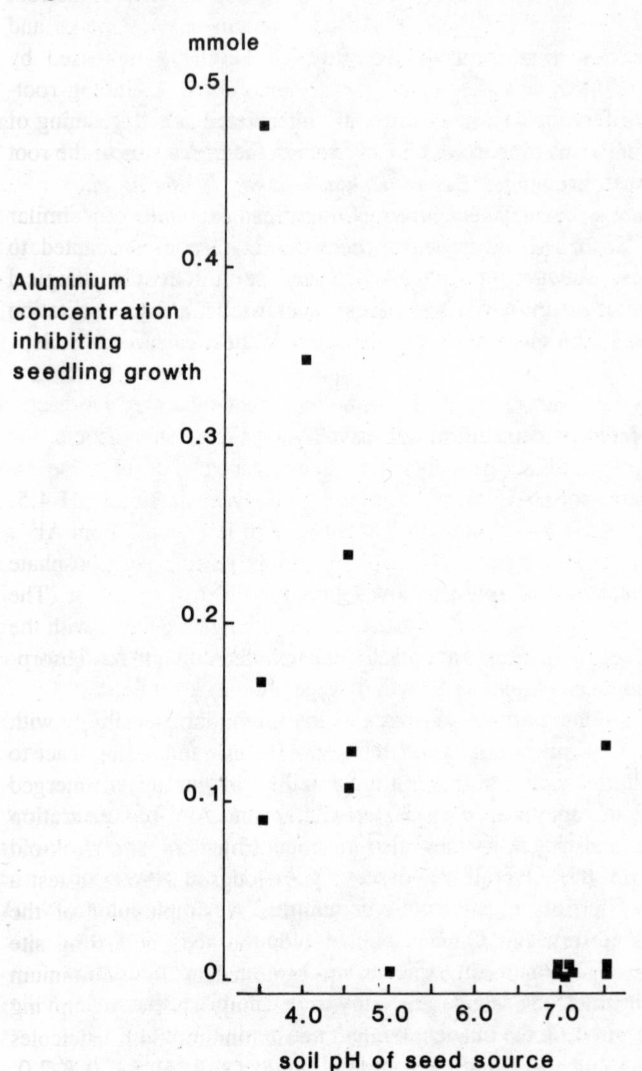

resolved when it is recognised that the soil peat, although acid and base poor, is almost totally free from aluminium (Sparling 1967).

Are calcicoles adapted to drought?

All the forms of adaptation to water shortage noted generally in Chapter 3 may be found in calcicoles, e.g. deep-rooted microphyllous perennials; *Helianthemum canum, Juniperus phoenicea, Thymus* spp.; succulent water storage; *Sempervivum* spp.; annual herb: *Saxifraga tridactylites*; deep fibrous or tap-rooted forb: *Cirsium acaulon*; geophytes: *Allium* spp., Orchidaceae, e.g. *Epipactis atrorubens*; desiccation tolerance: *Ceterach officinale, Ramonda* spp.

However, all that this list indicates is that dry, calcareous subhabitats frequently occur. In upland areas they may represent a very high percentage of all calcareous sites. This should not, however, lead to the conclusion that independence of these two factors is not possible.

Are calcicoles adapted to low fertility?

Calcareous situations have been detected where major nutrients are certainly a constraint to growth of vegetation as a whole, e.g. Willis (1963), dune pasture, nitrogen; Lloyd and Pigott (1967), chalk grassland, nitrogen; and Jeffrey (1971), upland grassland, phosphorus. Some vegetation types fail to respond to nitrogen and phosphate addition, e.g. *Sesleria* grassland (Jeffrey and Pigott 1973). Individual species also respond, e.g. *Lathyrus pratensis* (Grime 1965). Overall we possess a poor sense of proportion on the place of fertility in calcicole communities. Our own ongoing work (D.W. Jeffrey and G. O'Donovan, unpublished), is indicating a picture of slow nutrient turnover and low primary production in the Burren limestone vegetation. However, more comparative studies are required on the nutrient content and availability of a wide range of limestone soils; and on community nutrient turnover. It may be that soils and vegetation types will be ranked according to the chemical property of the parent material. The massive pale Carboniferous limestones could well have lower phosphate, potassium and micronutrient contents compared with darker, softer and younger limestone series. Sand-dune soils are well supplied with phosphorus, and provide good conditions for symbiotic and free-living nitrogen-fixing types. Dryness is probably the prime limitation to nitrogen fixation and turnover in calcareous dunes. The alpine series of Ozenda (1983) would make a most interesting series to compare in fertility terms. Finally, mention should be made of wet calcareous communities, especially the wet fen types. There is some evidence suggesting a soil phosphorus gradient between the various subunits.

To explore fully the comparative fertility of calcareous soils, more work needs to be done in developing availability indices. The role of mycorrhizas and other microbial populations in assimilation and nutrient turnover also needs attention.

SERPENTINE SOILS

In geological contrast to the sedimentary limestones, serpentine rocks are derived from exposure of igneous intrusions of a wide range of ferromagnesian minerals to hydrothermal activity. A simple example is the conversion of olivine to serpentine by the action of CO_2-charged hot water:

$$\underset{\text{Olivine}}{4Mg_2SiO_4} + 4H_2O + 2CO_2 \rightarrow \underset{\text{Serpentine}}{Mg_6Si_4O_{10}(OH)_8} + \underset{\text{Magnesite}}{2MgCO_3}$$

In this group of serpentine minerals, magnesium may be replaced by iron, nickel, cobalt or chromium. Another group contains variable quantities of calcium.

To sum up a variable and complex situation, serpentines are parent materials in which:

(a) Fe and Mg are high relative to Si;
(b) Ca is low relative to Mg;
(c) P, K and Mo may also be low;
(d) Ni, Cr and Co may be present in high, even toxic, amounts (Proctor and Woodell 1975).

A wide range of soil types arise from serpentines, dependent on details of parent material, weathering regime, relief and biological processes. Many serpentine soils are shallow and stony, whereas others are deep with poor internal drainage. Physical properties must be added to the litany of possible factors unfavourable to plant growth. The fascinating feature of serpentine is that although correlated geochemically, the biological effects may be either independent or interlinked.

SERPENTINE VEGETATION

In all parts of the world where serpentine outcrops occur, a distinct vegetation variant is correlated with it. The vegetation is typically more open and apparently stunted. Three growth forms predominate: coniferous trees, sclerophyllous shrubs, and sclerophyllous forbs and graminoids. There is generally a high incidence of endemism; for example, the Californian crucifer genus *Streptanthus* has 16 taxa in one subgenus, nearly all of which are restricted to serpentine (Kruckeberg 1984). Total species number may be higher than in adjacent closed vegetation on other rock types. Serpentine ecotypes have been isolated that appear to be selected from species common to adjacent vegetation. The examples in Table 19.3 indicate the range of transitions that are observed in all climatic belts and altitudes.

Table 19.3: Serpentine vegetation types

Location	Serpentine	Adjacent non-serpentine	Author
Siskiyou Mtns, Oregon	Open woodland with isolated trees, a more xeric pine, evergreen shrubs and rare herb species	Closed dense evergreen forest with evergreen ground flora	Whittaker (1960)
Coast ranges, California, USA	(a) Serpentine chaparral of sclerophyllous shrubs (b) grassland Both include endemics and local indicator species	Oak–pine woodland	Kruckeberg (1984)
Quebec	Tundra	Taiga	Whittaker (1975)
Cuba and New Caledonia	Savannah scrub	Tropical rainforest	Whittaker (1975)
New Zealand	Tussock grassland	*Nothofagus* forest	Proctor and Woodell (1971)

SOME APPROACHES TO THE SERPENTINE PROBLEM

The overview approach

A series of isolated studies of serpentine, mostly from particular viewpoints, constitute the literature. The four groups of inhibitory factors listed in Table 19.4, namely physical factors, fertility factors, the Ca–Mg complex and the Ni–Cr–Co toxicity complex, each need separating. The problem with the background literature as listed by Proctor and Woodell (1975) is that a standard checklist for all four factor groups is missing. It is thus not possible to obtain a type classification that would separate substrate and vegetation on the particular combination of adverse factors present, and on the intensity of each adverse factor.

Table 19.4: Factors in serpentine soil environment

Physical characteristics
Almost invariably unfavourable, ranging from shallow and overdrained to deep and ill-drained and poorly aerated. Chemical weathering seems to dominate over biological effects in soil development.

Limiting supply of nitrogen, phosphate and potassium
Composition of plants and response to added nutrients are variable. A depressant effect of chromium on phosphate availability seems possible in the rare Cr-rich serpentines.

Calcium–magnesium relationships
Exchangeable Mg is almost invariably high and exchangeable Ca low. Three distinct effects seem possible:

(a) Unfavourable Ca : Mg ratios. These are typically in the range 0.1–0.4, whereas ratios for non-serpentine soils are greater than 1.0 (Proctor and Woodell 1975).
(b) Ca deficiency. In a number of experiments, Ca addition has improved growth where N, P and K did not. Absorption of Ca leads to a better plant Ca : Mg ratio in some cases. Ca may also have a protective role in the face of Ni and Cr toxicity.
(c) Mg toxicity. Where tested, most serpentine endemic species are tolerant to high Mg concentrations, whereas 10 ppm Mg in solution is seriously toxic to non-serpentine types.

Toxicity of other metals
Geological and weathering variables frequently lead to substrates with anomalously high concentrations of Ni, Cr and Co. There is a large literature dealing with geochemistry, availability, absorbance and accumulation by plants, physiological interactions and the tolerance to these metals of serpentine ecotypes. These are reviewed in detail by Proctor and Woodell (1975) and in a broader context by Woolhouse (1983). Nickel toxicity and its response in terms of tolerance is most frequent and of greatest ecological importance. Chromium is an occasional problem, anomalous cobalt seems to have little ecological importance.

Table 19.5: A hypothetical classification of serpentine situations based on the occurrence of four adverse factor groups

	Low Ca : Mg ratio	Toxic metal anomaly	Low soil fertility	Adverse physical conditions
Single factors				
1	+			
2		+		
3			+	
4				+
Double factors				
5	+	+		
6	+		+	
7	+			+
8		+	+	
9		+		+
10			+	+
Triple factors				
11	+	+	+	
12	+	+		+
13	+		+	+
14		+	+	+
Quadruple factors				
15	+	+	+	+

To have a such a world-wide perspective would place all ongoing studies in better context. The checklist might pose four simple questions about each serpentine locality to give a first approximation:

(1) Is there evidence from soil, vegetation, or both, that a low Ca : Mg ratio exists?
(2) Do anomalously high concentrations of any other element occur, Ni, Cr or Co in particular?
(3) Do nutrient availability studies or any other evidence point to nutrient deficiency?
(4) Does soil physical structure suggest adverse conditions greater than experienced by adjacent non-serpentine vegetation?

A yes or no answer would give rise in theory to fifteen possible situations (Table 19.5). It would be helpful to eliminate as many possibilities as possible, and to indicate sites of value in solving special parts of the overall problem. For example, are there any serpentine sites in which only one adverse factor operates?

The analytic approach

For an individual site or group of sites, an approach similar to that employed for mine wastes can be used (see Chapter 14). This could take the form of the following investigations:

(a) Evaluation of substrate according to the above checklist.
(b) Inventory of vegetation, combined with selected chemical analyses; i.e. do the growth form, species make-up, and chemical compositions of vegetation relate to the overall literature?
(c) Select test species or possible ecotypes from serpentine and adjacent non-serpentine vegetation. These might be species readily propagated vegetatively.
(d) Design a series of experiments based on the above information in which hypothetical limiting factors are removed individually, using serpentine and non-serpentine test plants and a suitable control substrate.

It must be remembered, however, that in this situation multiple depressant factors are assumed to be operating. Failure to respond to, say, an irrigation regime by a non-serpentine species is by no means proof that water supply is non-limiting. The use of serpentine and non-serpentine types is thus of great importance. A single factor may trigger a response in a serpentine endemic or ecotype rather than in a non-serpentine plant. This would indicate the effects of adverse-factor amelioration added to the effects of tolerance to other adverse factors. For example:

(i) Water stress: transfer substrate to suitable containers and maintain under field moisture and under an irrigated regime.
(ii) Soil fertility could be investigated by field experiments and by pot experiments in which water relationships can also be controlled.
(iii) The Ca–Mg complex should be approached first through calcium additions, as solution and as a variety of solid-phase sources, e.g. gypsum ($CaSO_4$), hydrated lime ($Ca(OH)_2$), limestone ($CaCO_3$). The combination of irrigation, NPK and Ca addition should be employed.
(iv) Toxic levels of nickel, chromium and cobalt can be determined by plant analysis, by multielement analysis of soil using X-ray fluorescence, and by availability studies using ammonium

acetate or dilute acid as extractants. Two experimental approaches can be employed to reduced toxicity. The substate (< 2 mm sieved) can be diluted using an inert material such as silica sand. Alternatively, a source of organic matter, peat or paper pulp, can be mixed with the substrate. The plant response should be evaluated as dry weight increment and as metal content (see Chapter 14).

A suite of experiments of this kind will test the inhibitory power of each factor group in a rather crude and preliminary way. The results, however, will permit more precise experiments to be designed. For example, assume that adjusting the Ca : Mg ratio using $CaSO_4$ and adding NPK to an irrigated system gave a good increase in growth of a serpentine plant. Furthermore a non-serpentine species at least established itself on the serpentine. This result would indicate the non-intervention of a factor such as nickel toxicity. This would permit experiments on (a) the fertility complex, i.e. is it a nitrogen, phosphate or potassium effect? (b) the Ca–Mg complex, checking the three possibilities set out in Table 19.3; and would determine whether (c) the response to irrigation is independent of other factors or whether it is interlinked with, say, rooting depth or soil organic matter.

Progressive refinement of hypotheses will lead inevitably to ecophysiological considerations which are too wide-ranging for concise speculation.

An approach via standardised or synthetic substrates

The variability of substrates in the field may well prove to be of great value when properly understood. However, it is a hindrance to comparative experimentation. Some benefits may be gained by exploring serpentine plant behaviour on carefully selected substrates. These may be available in particular field sites or alternatively it may be possible to synthesise substrates based on known ferromagnesium silicates. It would be possible to explore the Ca : Mg ratio situation and the relationship between Ni and the Ca–Mg complex in a way not possible with single substrates. We have used this approach to study multielement toxicities in mine wastes (Jeffrey and Maybury 1981).

Table 19.6: Commentary on Figure 19.3. Hypothetical interactions on serpentine soils

(1) The geochemical control affects plant growth and other biological activities directly, thus influencing soil formation indirectly. Organic matter input, and hence cation exchange properties, would be affected. Nickel, low calcium and high magnesium effects are depicted as acting independently, but low calcium may enhance the effect of nickel toxicity. Toxic effects will also be greater in the absence of substantial organic matter.
(2) Availablity of nutrients is affected by bedrock composition and by all the biological activities contributing to mineralisation and soil formation.
(3) Nitrogen supply, in terms of nitrogen fixation, turnover, and mineralisation, may be restricted by adverse geochemistry, and soil–water relationships, i.e. drought made worse by coarse particle size, low organic matter and shallow depth.
(4) Soil development, arrested by low organic input, may be further inhibited by topographic factors and low rainfall in some sites, e.g. in California.
(5) All biological activities may be limited by nutrients, toxic substances and low soil moisture.
(6) Water availability depends on the soil water storage capacity. This is limited as in (3) above.

Figure 19.3: Hypothetical multiple interactions which may govern plant growth and soil development on serpentine substrates

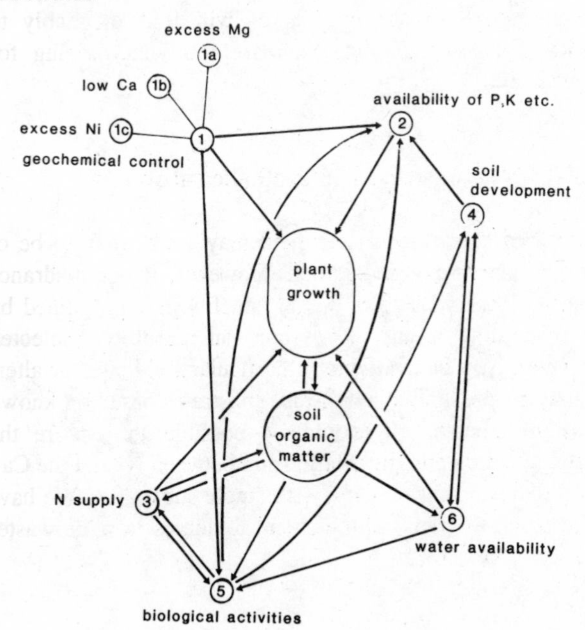

Using a model as a thinking tool

Diagrams of multiple interactions are frequently helpful in contemplating active investigation. This is especially true in this case, when a large number of possibilities for experimentation present themselves. The classical mode of single-factor experimentation may not be entirely suitable when chains of interaction occur. The model in Table 19.6 and Figure 19.3 depicts a tightly locked system. Growth constraints are imposed on the plants directly via the toxic geochemical factors and indirectly through their effect on microbial and other soil activities. Soil organic matter, with possible ameliorating properties, is dependent on plant growth. Biological input to weathering and soil development, in particular generation of soil aggregates and weathering depth, are constrained by all the interactions.

Further Reading

(Refer to chapters as indicated)

Alexander M. (1977) *Introduction to soil microbiology*, John Wiley, New York (Chs. 4 & 5)

Allen S.E., Grimshaw H.M., Parkinson J.A. and Quarmby C. (1974) *Chemical analysis of ecological materials*, Blackwell, Oxford (Ch. 12)

Barer L.D., Gardner W.H. and Gardner W.R. (1972) *Soil physics* 4th Edn, John Wiley, New York (Chs. 8 & 9)

Beeftink W.G., Rozema J. and Huiskes A.H.L. (1985) *Ecology of coastal vegetation*, Dr W. Junk, Dordrecht (Ch. 18)

Bewley J.D. (1979) Physiological aspects of desiccation tolerance. *Annual Review of Plant Physiology, 30*, 195–238 (Ch. 3)

Bielski R.B. (1973) Phosphate pools, phosphate transport and phosphate availability. *Annual Review of Plant Physiology, 24*, 225–52 (Ch. 2)

Bolin B. and Cook R.B. (Eds), (1983) *Scope 21. The major biogeochemical cycles and their interactions*, John Wiley, Chichester (Ch. 2)

Bormann F.H. and Likens G.E. (1979) *Pattern and process in a forested ecosystem*, Springer, New York (Ch. 15)

Bowen G.D. and Rovira A.D. (1971) Relationship between root morphology & nutrient uptake. In *Recent advances in plant nutrition*, pp. 293–306, Ed. R.M. Samish, Gordon & Breach Science Publishers, New York (Ch. 1)

Brooks R.R. (1987) *Serpentine and its vegetation*, Croom Helm, London/ Dioscorides Press, Portland, Oregon

Chabot B.F. and Mooney H.A. (1985) *Physiological ecology of North American plant communities*, Chapman & Hall, New York (Chs. 3, 17, 18, 19)

Chapman V.J. (Ed.) (1977) *Wet coastal ecosystems*, Elsevier, Amsterdam (Ch. 18)

Coleman D.C., Reid C.P.P. and Cole C.V. (1983) Biological strategies of nutrient cycling in soil systems. *Advances in Ecological Research, 13* 1–57 (Chs. 4, 5)

Cooper-Driver G.A., Swain T. and Conn E.E. (eds) (1985) Chemically mediated interactions between plants and other organisms. *Recent advances in phytochemistry, 19*, Plenum Press, New York (Chs. 4, 5)

Cosgrove D.J. (1977) Microbial transformations in the phosphorus cycle. *Advances in Microbial Ecology, 1*, 95–134 (Ch. 10)

Dalal R.C. (1977) Soil organic phosphorus. *Advances in Agronomy, 29*, 83–117 (Ch. 10)

Darwin C. (1875) *Insectivorous plants*, Murray, London (Ch. 5)

Dobereiner J. and De-Polli H. (1980) Diazotrophic Rhizocoenoses. In *Nitrogen fixation* pp. 30–34, Eds W.D.P. Stewart and J.R. Gallon, Annual Proceedings of the Phytochemical Society of Europe No. 18, Academic Press, London (Ch. 4)

Dommergues Y.R. and Krupa S.W. (Eds) (1970) *Interactions between non-pathogenic soil micro-organisms and plants*, Elsevier, Amsterdam (Chs. 4, 5)

Fitter A.H., Atkinson D. and Read D.J. (1985) *Ecological interactions in soil, plants, microbes and animals*, Blackwell Scientific Publications, Oxford (Chs. 4, 5)

Gibson A.H. and Jordon D.C. (1983) Ecophysiology of nitrogen-fixing systems. In *Encyclopedia of Plant Physiology III* pp. 302–20, Eds. O.L. Lange P.S. Nobel C.B. Osmond and H. Ziegler Springer-Verlag, Berlin (Chs. 4, 10)

Goodall D.W., Perry R.A. and Howes K.M.W. (1979–81) *Arid-zone ecosystems: Structure functioning and management*, Vols 1 and 2, Cambridge University Press, Cambridge (Ch. 3)

Greenland D.J. and Hayes M.H.B. (Eds) (1981) *The chemistry of soil processes*, John Wiley, Chichester (Chs. 10, 11)

Hanson A.D. and Hitz W.D. (1982) Metabolic responses of mesophytes to soil water deficits. *Annual Review of Plant Physiology, 33*, 163–203 (Ch. 3)

Heslop-Harrison Y. (1978) Carnivorous plants. *Scientific American, 238*, 104–15 (Ch. 5)

Hesse P.R. (1971) *A textbook of soil chemical analysis*, John Murray, London (Ch. 12)

Iler R.K. (1979) *The chemistry of silica*, John Wiley, New York (Ch. 2)

Kozlowski T.T. (Ed.) (1984) *Flooding and plant growth*, Academic Press, Orlando (Ch. 8)

Lamont B.B., Brown G. and Mitchell D.T. (1984) Structure, environmental effect on their formation, and function of Proteoid roots in *Leucodendron laureolum* (Proteaceae). *The New Phytologist, 97*, 381–90 (Ch. 16)

Lange O.L., Kappen L. and Schalze E.-D. (Eds) (1976) *Water and plant life*, Springer, Berlin (Ch. 3)

Lange O.L., Nobel P.S., Osmond C.B. and Ziegler H. (Eds) (1983) *Encyclopedia of plant physiology* New Series, Vol. 12C Physiological plant ecology III, Responses to the chemical and biological environment. Vol 12D Physiological plant ecology IV, Ecosystem processes: Mineral cycling, Productivity and man's influence. Springer, Berlin (Chs. 1, 2, 4, 10, 11, 14, 18, 19)

Loneragan J.F., Robson A.D. and Graham R.D. (Eds) (1981) *Copper in soils and plants*, Academic Press, Sydney (Chs. 10, 14)

Marks G.C. and Kozlowski T.T. (1973) *Ectomycorrhizae*, Academic Press, New York (Ch. 4)

Martin F., Marchal J.-P., Timinsk A. and Canet D. (1985) The metabolism and physical state of polyphosphates in ecto mycorrhizal fungi. A ^{31}P nuclear magnetic resonance study. *The New Phytologist, 101*, 275–90 (Ch. 4)

Milewski A.V. (1983) A comparison of ecosystems in Mediterranean Australia and southern Africa. *Annual Review of Ecology and Systematics, 14*, 57–76 (Chs. 3, 16)

Mosse B., Stribley D.P. and Le Tacon F. (1981) Ecology of mycorrhizae and mycorrhizal fungi. *Advances in Microbial Ecology, 5*, 137–210 (Ch. 4)

Nye P.H. and Tinker P.B. (1977) *Solute movement in the soil-root system*,

Blackwell, Oxford (Chs. 1, 2, 4, 10, 11)
Parker G.G. (1983) Throughflow and stemflow in the forest nutrient cycle. *Advances in Ecological Research, 13*, 58–135 (Ch. 15)
Pate J.J. (1980) Transport and partitioning of nitrogenous solutes. *Annual Review of Plant Physiology, 31*, 313–40 (Chs. 2, 11)
Phillips D.A. (1980) Efficiency of symbiotic nitrogen fixation in legumes. *Annual Review of Plant Physiology, 31*, 29–49 (Chs. 4, 11)
Postgate J.R. (1982) *The fundamentals of nitrogen fixation*, Cambridge University Press, Cambridge (Chs. 4, 11)
Robb D.A. and Pierpoint W.S. (Eds) (1983) Metals and micronutrients: Uptake and utilization by plants. *Annual Proceedings of the Phytochemical Society of Europe*, No. 21, Academic Press, London (Chs. 2, 10, 11, 14, 18)
Savont N.K. and De Datta S.K. (1982) Nitrogen transformations in wetland rice soils. *Advances in Agronomy, 35*, 241–302 (Ch. 4)
Sprent J. (1979) *The biology of nitrogen fixing organisms*, McGraw-Hill, London (Ch. 4)
Stewart W.D.P. and Gallon J.R. (Eds) (1980). Nitrogen fixation. *Annual Proceedings of the Phytochemical Society of Europe*, No. 18. Academic Press, London (Chs. 4, 10)
Straker C.J. and Mitchell D.T. (1985) The characterisation and estimation of polyphosphates in endomycorrhizas of the Ericaceae. *New Phytologist, 99*, 341–440 (Chs. 4, 16)
Subba Rao N.S. (1980) *Recent advances in biological nitrogen fixation*. Edward Arnold, London (Chs. 4, 10)
Swain T. (1977) Secondary compounds as protective agents. *Annual Review of Plant Physiology, 28*, 479–501 (Chs. 2, 5)
Swift M.J., Heal O.W. and Anderson J.M. (1979) *Decomposition in terrestrial ecosystems*. Blackwell, Oxford (Ch. 5)
Timmermann B.N., Steelink C. and Loewus F.A. (Eds) (1984) *Phytochemical adaptations to stress. Recent advances in Photochemistry*, vol. 18. Plenum Press, New York (Chs. 3, 14, 18, 19)
Walter H. (1973) *Vegetation of the Earth in relation to climate and ecophysiological conditions*. (English translation, J. Wieser), English Universities Press, London (Ch. 1)
Whipps J.M. and Lynch J.M. (1986) The influence of the rhizosphere on crop productivity. *Advances in Microbial Ecology, 9*, 187–244 (Ch. 4)
Wiebe H.H. (Ed.) (1971) *Measurement of plant and soil water status*. Utah Agricultural Experimental Station, Logan, Utah (Chs. 3, 8)

Bibliography

Adamson R.S. (1927) The plant communities of Table Mountain: preliminary account. *Journal of Ecology, 15*, 278–309

Akkermans A.D.L. and van Dijk C. (1981) Non-leguminous root-nodule symbioses with actimomycetes and *Rhizobium*. In *Nitrogen fixation, Vol. 1, Ecology*, pp. 57–103, Ed. W.J. Broughton, Clarendon Press, Oxford

Anderson J.M., Huish S.A., Ineson P., Leonard M.A. and Splatt P.R. (1985) Interactions of invertebrates, microorganisms and tree roots in nitrogen and mineral element fluxes in deciduous woodland soils. In *Ecological interactions in soil*, pp. 377–92, Eds A.H. Fitter, D. Atkinson, D.J. Read and M.B. Usher, Blackwell, Oxford

Antonovics J., Bradshaw A.D. and Turner R.G. (1971) Heavy metal tolerance in plants. *Advances in Ecological Research, 7*, 1–85

Armstrong W. (1982) Waterlogged soils. In *Environment and plant ecology*, pp. 290–327, J.R. Etherington, John Wiley, Chichester

Arnon D.I. and Stout P.R. (1939) The essentiality of certain elements in minute quantity for plants with special reference to copper. *Plant Physiology, 14*, 371–5

Barber S.A. (1984) *Soil nutrient bioavailability*, John Wiley, New York

Batzli G.O., White R.G., Maclean S.F., Pitelka F.A. and Collier B.D. (1980) The herbivore based trophic system. In *An Arctic ecosystem*, pp. 335–457, Eds J. Brown, P.C. Miller, L.L. Tieszen and F.L. Bunnell, Dowden, Hutchinson & Ross, Stroudsburg, PA

Baylis G.T.S. (1975) The magnolioid mycorrhiza and mycotrophy in root systems derived from it. In *Endomycorrhizas*, pp. 373–89, Eds F.E. Sanders, B. Mosse and P.B. Tinker, Academic Press, London

Beadle N.C.W. (1966) Soil phosphate and its role in moulding segments of the Australian flora and its vegetation, with special reference to xeromorphy and sclerophylly. *Ecology, 47*, 992–1007

Beeftink W.G. (1977) Saltmarshes. In *The coastline*, Ed. R.S.K. Barnes, John Wiley, London

Benzing D.H. (1973) Mineral nutrition and related phenomena in Bromeliaceae and Orchidaceae. *Quarterly Review of Biology, 48*, 277–90

Bernays E.A. (1983) Nitrogen in defence against insects. In *Nitrogen as an ecological factor*, pp. 321–44, Eds J.A. Lee, S. McNeill and I.H. Rorison, Blackwell, Oxford

Bowen G.D. (1973) Mineral nutrition of ectomycorrhizae. In *Ectomycorrhizae*, pp. 151–206, Eds G.C. Marks and T.T. Kozlowski, Academic Press, New York

Bowen H.J.M. (1979) *Environmental chemistry of the elements*, Academic Press, London

Bradshaw A.D. and Chadwick M.J. (1980) *The restoration of land*, Blackwell, Oxford

Braekke F.D. (Ed.) (1976) Impact of acid precipitation on forest and freshwater ecosystems in Norway. *Summary Report S.N.S.F. Project*, Oslo

Briens M. and Lahrer F. (1982) Osmoregulation in halophytic higher plants: a comparative study of soluble carbohydrates, polyols, betaines and free proline. *Plant, Cell and Environment, 5*, 287–92

Brooks R.R. (1979) Advances in botanical methods for prospecting for minerals. Part II. Advances in biogeochemical methods for prospecting. In *Geophysics and geochemistry in the search for metallic ores*, pp. 397–410, Ed. P.J. Hood, Geological Survey of Canada. Economic Geology Report 31, Ottawa

Brown J., Miller P.C., Tieszen L.L. and Bunnell F.L. (1980) *An Arctic ecosystem. The central tundra at Barrow, Alaska*, Dowden, Hutchinson & Ross, Stroudsburg, PA

Brown J.C. (1978) Mechanisms of iron uptake by plants. *Plant, Cell and Environment, 1*, 249–57

Bulow-Olsen A. (1980) Nutrient cycling in grassland dominated by *Deschampsia caespitosa* (L)Trin. and grazed by nursing cows. *Agro-Ecosystems, 6*, 209–20

Burges A. (1960) Time and size as factors in ecology. *Journal of Ecology, 48*, 273–85

Caldwell M.M. (1976) Root extension and water absorption. In *Water and plant life*, pp. 63–85, Eds O.L. Lange, L. Kappen and E.-D. Schulze, Springer, Berlin

Caruccio F.T. (1975) Estimating the acid potential of coal mine refuse. In *The ecology of resource degradation and renewal*, Eds M.J. Chadwick and G.T. Goodall, Blackwell, Oxford

Chandler R.J. (1942) The time required for podzol profile formation as evidenced by the Mendenhall glacial deposits near Juneau, Alaska. *Proceedings of the Soil Science Society of America, 7*, 454–9

Chapin F.S. III, Johnson D.A. and McKendrick J.D. (1980) Seasonal movements of nutrients in plants of differing growth forms in an Alaskan tundra ecosystem: implications for herbivory. *Journal of Ecology, 68*, 189–209

Chapin F.S. III., Miller P.C., Billings W.D. and Coyne P.I. (1980) Carbon and nutrient budgets and their control in coastal tundra. In *An Arctic Ecosystem*, pp. 458–82, Eds J. Brown, P.C. Miller, L.L. Tieszen and F.L. Bunnell, Dowden, Hutchinson & Ross, Stroudsburg, PA

Chapman V.J. (1960) *Saltmarshes and salt deserts of the world*. Hill, London

Clarkson D.T. (1969) Metabolic aspects of aluminium toxicity and some possible mechanisms for resistance. In *Ecological aspects of the mineral nutrition of plants*, pp. 381–97, Ed. I.H. Rorison, Blackwell, Oxford

Clarkson D.T. (1984) Ionic relations. In *Advanced plant physiology*, pp. 319–53, Ed. M.B. Wilkins, Pitman, London

Clarkson D.T. and Hanson J.B. (1980) The mineral nutrition of higher plants. *Annual Review of Plant Physiology, 31*, 239–98

Cole D.W. and Rapp M. (1981) Elemental cycling in forest ecosystems. In *Dynamic properties of forest ecosystems*, pp. 341–410, Ed. D.E. Reichle, Cambridge University Press, Cambridge

Corby H.D.L., Pohill R.M. and Sprent J.I. (1983) Taxonomy. In *Nitrogen fixation, Vol. 3. Legumes*, pp. 1–35, Ed. W.J. Broughton, Oxford University Press, Oxford

Cosgrave D.J. (1980) *Inositol phosphates: their chemistry, biochemistry and physiology*, Elsevier, Amsterdam

Coupland R.T. (Ed.) (1979) *Grassland ecosystems of the world. International Biographical Programme Vol. 18*, Cambridge University Press, Cambridge

Cox G., Saunders F.E., Tinker P.B. and Wild J.A. (1975) Ultrastructural evidence relating to host–endophyte transfer in a vesicular–arbuscular mycorrhiza. In *Endomycorrhizas*, pp. 297–312, Eds F.E. Saunders, B. Mosse and P.B. Tinker, Academic Press, London

Crawford R.M.M. (1978) Metabolic responses to anoxia. In *Plant life in anaerobic environments*, Eds D.D. Hook and R.M.M. Crawford, Ann Arbor Science, Michigan

Crocker R.C. (1952) Soil genesis and the pedogenic factors. *Quarterly Review of Biology, 27*, 139–68

Cronquist A. (1981) *An integrated system of classification of flowering plants*, Columbia University Press, New York

Daines R.J. and Gould A.R. (1985) The cellular basis of salt tolerance studied with tissue cultures of the halophytic grass *Distichlis spicata. Journal of Plant Physiology, 119*, 269–80

Daubenmire R. (1972) Ecology of *Hyparrhenia rufa* in derived savannah in north-western Costa Rica. *Journal of Applied Ecology, 9*, 11–23

Dieter P. (1984) Calmodulin and calmodulin-modulated processes in plants. *Plant, Cell and Environment, 7*, 371–80

Dunn E.L., Shropshire F.M., Song L.C. and Mooney H.A. (1976) The water factor and convergent evolution in Mediterranean-type vegetation. In *Water and plant life*, pp. 492–505, Eds O.L. Lange, L. Kappen and E.-D. Schulze, Springer, Berlin

Ente P.J. (1967) Initial decalcification due to oxidation of sulphides in young marine soils in the Netherlands. *Transactions VIII. International Congress of Soil Science, 2*, 779–83

Feldman L.J. (1984) Regulation of root development. *Annual Review of Plant Physiology, 35*, 223–42

Flemming G.A. (1973) Mineral composition of herbage. In *Chemistry and biochemistry of herbage*, Vol. 1, pp. 529–66, Eds G.W. Butler and R.W. Bailey, Academic Press, London

Foth H.D. (1978) *Fundamentals of soil science*. 6th Edn, John Wiley, New York

Gasser J.K.R. (1982) Agricultural productivity and the nitrogen cycle. *Philosophical Transactions of the Royal Society of London, Series B, 296*, 303–14

Gates D.M. (1962) *Energy exchange in the biosphere*, Harper & Row, New York

Geiger R. (1965) *The climate near the ground*, Harvard University Press, Cambridge, Mass.

Gerrard A.J. (1981) *Soils and landforms*, George Allen & Unwin, London

Gill A.M. (1975) Fire and the Australian flora, a review. *Australian Forestry, 38*, 4–25

Giller K.E. and Day J.M. (1985) Nitrogen fixation in the rhizosphere: significance in natural and agricultural systems. In *Ecological interactions in soil*, pp. 127–48, Eds A.H. Fitter, D. Atkinson, D.J. Read

and M.B. Usher, Blackwell, Oxford
Gimingham C.H. (1972) *Ecology of heathlands*, Chapman & Hall, London
Girnish T.J., Burkhardt E.L., Happel R.E. and Weintraub J.D. (1984) Carnivory in the bromeliad *Brocchinia reducta*, with a cost/benefit model for the general restriction of carnivorous plants to sunny, moist, nutrient-poor habitats. *The American Naturalist, 124*, 479–97
Glass A.D.M. (1983) Regulation of ion transport. *Annual Review of Plant Physiology, 34*, 311–26
Goodman P.J. and Williams W.T. (1961) Investigations into 'die-back' in *Spartina townsendii* agg. III Physiological correlates of die-back. *Journal of Ecology, 49*, 391–8
Gorham J., Hughes L. and Wyn-Jones R.H. (1980) Chemical composition of saltmarsh plants from Ynys Mon (Anglesey): the concept of physiotypes. *Plant, Cell and Environment, 3*, 309–18
Griffin D.M. (1972) *Ecology of soil fungi*, Chapman & Hall, London
Grime J.P. (1963) Factors determining the occurrence of calcifuge species on shallow soils over calcareous substrata. *Journal of Ecology, 51*, 375–90
Grime J.P. (1965) The ecological significance of lime chlorosis (an experiment with two species of *Lathyrus*). *New Phytologist, 64*, 477–88
Grime J.P. and Hodgson J.G. (1969) An investigation of the ecological significance of lime chlorosis by means of large-scale comparative experiments. In *Ecological aspects of the mineral nutrition of plants*, Ed. I.H. Rorison, Blackwell, Oxford
Groves R.H. (1964) *Experimental studies on heath vegetation*. PhD Thesis, University of Melbourne
Groves R.H. and Specht R.L. (1965) Growth of heath vegetation. I. Annual growth curves on two heath ecosystems in Australia. *Australian Journal of Botany, 13*, 261–80
Hardy R.W.F., Burns R.C., Hebert R.R., Holsten R.D. and Jackson E.K. (1971) Biological nitrogen fixation: a key to world protein. In *Biological nitrogen fixation in natural and agricultural habitats. Plant and Soil Special Volume*, pp. 561–90, Eds T.A. Lie and E.G. Mulder
Hardy R.W.F., Holsten R.D., Jackson E.K. and Burns R.C. (1968) The acetylene–ethylene assay for N_2 fixation: laboratory and field evaluation. *Plant Physiology, 43*, 1185–1207
Harold F.M. (1966) Inorganic polyphosphates in biology: structure, metabolism and function. *Bacteriological Reviews, 30*, 772–4
Harper S. (Transl.) (1982) *Acidification today and tomorrow*, Ministry of Agriculture (Sweden), 1982 Committee, Stockholm
Harrison A.F. and Bocock K.L. (1981) Estimation of soil bulk-density from loss-on-ignition values. *Journal of Applied Ecology, 8*, 919–27
Havill D.C., Ingold A. and Pearson J. (1985) Sulphide tolerance in coastal halophytes. *Vegetatio, 62*, 279–85
Heinrich B. and Bartholomew G.A. (1979) The ecology of the African dung beetle. *Scientific American, 241*, 118–26
Hewitt E.J. and Smith T.A. (1975) *Plant mineral nutrition*, English Universities Press, London
Hillel D. (1980) *Fundamentals of soil physics*, Academic Press, New York
Hutchinson J. (1973) *The families of flowering plants*, 3rd Edition, Oxford

University Press, Oxford

Jackson M.J. and Drew M.C. (1984) Effects of flooding on growth and metabolism of herbaceous plants. In *Flooding and plant growth*, pp. 47–128, Ed. T.T. Kozlowski, Academic Press, Orlando, FL

Jefferies R.L. (1977) Growth responses of coastal halophytes to inorganic nitrogen. *Journal of Ecology, 65*, 847–66

Jefferies R.L. (1980) Organic solutes in osmotic regulation in halophytic plants. In *Genetic engineering of osmotic regulation*, Eds D.W. Rains, R.C. Valentine and A. Hollaender, Plenum Press, London

Jefferies R.L. and Perkins N. (1977) The effects on the vegetation of the additions of inorganic nutrients to saltmarsh soils at Stiffkey, Norfolk. *Journal of Ecology, 65*, 867–82

Jeffrey D.W. (1964) The formation of polyphosphate in *Banksia ornata*, an Australian heath plant. *Australian Journal of Biological Science, 17*, 845–54

Jeffrey D.W. (1967) Phosphate nutrition of Australian heath plants. I. The importance of proteoid roots in *Banksia* (Proteaceae). *Australian Journal of Botany, 15*, 403–11

Jeffrey D.W. (1968) Phosphate nutrition of Australian heath plants. II. The formation of polyphosphate by five heath species. *Australian Journal of Botany, 16*, 603–13

Jeffrey D.W. (1970) A note on the use of ignition loss as a means for the approximate estimation of soil bulk density. *Journal of Ecology, 58*, 297–9

Jeffrey D.W. (1971) The experimental alteration of a *Kobresia*-rich sward in upper Teesdale. In *The scientific management of animal and plant communities for conservation*, Eds E. Duffey and A.S. Watt, Blackwell, Oxford

Jeffrey D.W. and Maybury M.M. (1981) Scientific studies of mine waste revegetation: the assessment and reduction of heavy metal toxicity in the revegetation of mining wastes. *Irish Journal of Environmental Science, 1*, 49–56

Jeffrey D.W. and Piggott C.D. (1973) The response of grasslands on sugar-limestone in Teesdale to applications of phosphorus and nitrogen. *Journal of Ecology, 61*, 85–92

Jeffrey D.W., Goodwillie R.N., Healy B., Holland C.H. and Moore J.J. (Eds) (1977) *North Bull Island, Dublin Bay*, Royal Dublin Society, Dublin

Jeffrey D.W., Maybury M. and Levinge D. (1975) Ecological approach to mining waste revegetation. In *Minerals and the environment*, pp. 371–85, Ed. M.J. Jones, Institution of Mining and Metallurgy, London

Jenkinson D.S. (1981) The fate of plant and animal residues in soil. In *The chemistry of soil processes*, pp. 505–62, Eds D.J. Greenland and M.H.B. Hayes, John Wiley, Chichester

Jenny H. (1941) *Factors of soil formation*, McGraw Hill, New York

Jenny H. (1961) A derivation of state factor equations of soils and ecosystems. *Proceedings of the Soil Science Society of America, 25*, 385–8

Jenny H. (1962) Model of rising nitrogen profile in Nile valley alluvium, and its agronomic and pedogenic implications. *Proceedings of the Soil*

Science Society of America, 26, 588–91
Jones K. (1974) Nitrogen fixation in a salt marsh. *Journal of Ecology, 62*, 553–65
Kinzel H. (1983) Influence of limestone, silicates and soil pH on vegetation. In *Physiological plant ecology III. Responses to the chemical and biological environment*, pp. 201–44, Eds O.L. Lange, P.S. Nobel, C.B. Osmond and H. Zeigler, Springer, Berlin
Kirkby E.A. and Pilbeam D.J. (1984) Calcium as a plant nutrient. *Plant, Cell and Environment, 7*, 397–405
Kozlowski T.T. and Ahlgren C.E. (1974) *Fire and ecosystems*, Academic Press, New York
Kramer P.J. (1983) *Water relations of plants*, Academic Press, New York
Kruckeberg A.R. (1984) *California serpentines*, University of California Press, Berkeley, CA
Larsen S. (1967) Soil phosphorus. *Advances in Agronomy, 19*, 151–210
Lee J.A. and Woolhouse H.W. (1969) A comparative study of bicarbonate inhibition of root growth in calcicole and calcifuge grasses. *New Phytologist, 68*, 1–11
Levinge D.E.S. (1977) *Nitrogen nutrition and mine waste revegetation*, PhD Thesis, University of Dublin, Trinity College
Likens G.E. (1985) The fifth Tansley lecture. An experimental approach for the study of ecosystems. *Journal of Ecology, 73*, 381–96
Likens G.E., Borman F.H., Pierce R.S., Eaton J.S. and Johnson N.M. (1977) *Biogeochemistry of a forested ecosystem*, Springer, New York
Likens P.C. (1984) *Publications of the Hubbard Brook Ecosystem Study*, Institute of Ecosystems Studies, Millbrook, New York
Lloyd P.S. and Pigott C.D. (1967) The influence of soil conditions on the course of succession on the chalk of southern England. *Journal of Ecology, 55*, 137–46
Loneragan J.F., Robson A.D. and Graham R.D. (Eds) (1981) *Copper in soils and plants*, Academic Press, Sydney
Loughman B.C. and Ratcliffe R.G. (1984) Nuclear magnetic resonance and the study of plants. *Advances in Plant Nutrition, 6*, 241–83
Lucas G. and Synge H. (1978) *The IUCN plant red data book*, IUCN, Morges
Luttge U. (1971) Structure and function of plant glands. *Annual Review of Plant Physiology, 22*, 23–44
Maarel E. van der (1971) Plant species diversity in relation to management. In *The scientific management of animal and plant communities for conservation*, pp. 45–63, Eds E. Duffey and A.S. Watt, Blackwell, Oxford
McBride M.B. (1981) Forms and distribution of copper in solid and solution phases of soil. In *Copper in soils and plants*, pp. 25–46, Eds J.F. Loneragan, A.D. Robson and R.D. Graham, Academic Press, Sydney
McIntyre D.S. (1970) The platinum electrode method for soil aeration measurement. *Advances in Agronomy, 22*, 235–83
McNamee K. (1976) *An experimental study of an Irish saltmarsh*, PhD Thesis, University of Dublin, Trinity College
McNamee K.A. and Jeffrey D.W. (1977) Ecophysiology of saltmarsh plants. In *North Bull Island, Dublin Bay*, pp. 100–6, Eds D.W. Jeffrey,

R.N. Goodwillie, B. Healy, C.H. Holland and J.J. Moore, Royal Dublin Society, Dublin

Magistad O.C. (1925) The aluminium content of the soil solution and its relation to soil reaction and plant growth. *Soil Science, 20*, 181–226

Martin M.H. (1968) Conditions affecting the distribution of *Mercurialis perennis* in certain Cambridgeshire woodlands. *Journal of Ecology, 56*, 777–93

Mattson W.J. (1980) Herbivory in relation to plant nitrogen content. *Annual Review of Ecology and Systematics, 11*, 119–61

Meidner H. and Sheriff D.W. (1976) *Water and plants*, Blackie, Glasgow

Milner C. and Hughes, R.E. (1968) *Methods for the measurement of primary productivity of grasslands*, Blackwell, Oxford

Moore P.D. and Bellamy D.J. (1973) *Peatlands*, Elek Science, London

Nassery H. and Harley J.L. (1969) Phosphate absorption by plants from habitats of different phosphate status. I. Absorption and incorporation of phosphate by excised roots. *New Phytologist, 68*, 13–20

Nedwell D.B. (1982) Exchange of nitrate, and the products of bacterial nitrate reduction between seawater and sediment from a UK saltmarsh. *Estuarine, Coastal and Shelf Science, 14*, 557–66

Newbould P.J. (1967) *Methods for estimating the primary production of forests*, Blackwell, Oxford

Nriagau J. (Ed.) (1979) *Copper in the environment*, John Wiley, New York

Olson, J.S. (1958) Rates of succession and soil changes on southern Lake Michigan sand dunes. *Botanical Gazette, 119*, 125–70

Ovington J.D. (1960) The afforestation of the Culbin sands. *Journal of Ecology, 38*, 303–19

Ozenda P.G. (1983) *The vegetation of the Alps*. Nature and Environment Series No. 29, Council of Europe, Strasbourg

Parker A.J. (1981) The chemistry of copper. In *Copper in soils and plants*, pp. 1–24, Eds J.F. Loneragan, A.D. Robson and R.D. Graham, Academic Press, Sydney

Pate J.S. (1983) Patterns of nitrate metabolism in higher plants and their ecological significance. In *Nitrogen as an ecological factor*, pp. 225–56, Eds J.A. Lee, S. McNeill and I.H. Rorison, Blackwell, Oxford

Peterson P.J. (1971) Unusual accumulations of elements by plants and animals. *Science Progress, 59*, 505–26

Phillips R. and Henshaw G.G. (1977) The regulation of synthesis of phenolics in stationary phase cell cultures of *Acer pseudoplatanus* L. *Journal of Experimental Botany, 28*, 785–94

Pigott C.D. (1956) The vegetation of upper Teesdale in the north Pennines. *Journal of Ecology, 44*, 545–86

Pigott C.D. (1969) Influence of mineral nutrition on the zonation of flowering plants in coastal saltmarshes. In *Ecological aspects of the mineral nutrition of plants*, pp. 25–35, Ed. I.H. Rorison, Blackwell, Oxford

Pigott C.D. and Taylor K. (1964) The distribution of some woodland herbs in relation to the supply of nitrogen and phosphorus in the soil. *Journal of Ecology, 52* (supplement), 175–85

Pitelka F.A. (1964) The nutrient-recovery hypothesis for arctic–microtine cycles. I. Introduction. In *Grazing in terrestrial and marine environments*, pp. 55–6, Ed. D.J. Crisp, Blackwell, Oxford

Pitman M.G. (1976) Ion uptake by plant roots. In *Encyclopedia of Plant Physiology*, Vol. 23, pp. 95–128, Eds N. Luttge and M.G. Pitman, Springer, Berlin

Poel L.W. (1960) The estimation of oxygen diffusion rates in soils. *Journal of Ecology, 48*, 165–73

Poole D.B.R., Moore L., Finch T.F., Gardiner M.J. and Flemming G.A. (1974) An unexpected occurrence of cobalt-pine in lambs in north Leinster. *Irish Journal of Agricultural Research, 13*, 119–22

Popp M. (1984a) Chemical composition of Australian mangroves. I. Inorganic ions and organic acids. *Zeitschrift für Pflanzenphysiologie, 113*, 395–409

Popp M. (1984b) Chemical composition of Australian mangroves. II. Low molecular weight carbohydrates. *Zeitschrift für Pflanzenphysiologie, 113*, 411–21

Popp M., Lahrer F. and Weigel P. (1984) Chemical composition of Australian mangroves. III. Free amino acids, total methylated onium compounds and total nitrogen. *Zeitschrift für Pflanzenphysiologie, 114*, 15–25

Proctor J. and Woodell S.R.J. (1971) The plant ecology of serpentine. I. Serpentine vegetation of England and Scotland. *Journal of Ecology, 59*, 375–95

Proctor J. and Woodell S.R.J. (1975) The ecology of serpentine soils. *Advances in Ecological Research, 9*, 256–366

Purnell H.M. (1960) Studies of the family Proteaceae. I. Anatomy and morphology of the roots of some Victorian species. *Australian Journal of Botany, 8*, 38–50

Ranwell D.S. (1972) *Ecology of salt marshes and sand dunes*, Chapman & Hall, London

Reichle D.E. (Ed.) (1981) *Dynamic properties of forest ecosystems*, Cambridge University Press, Cambridge

Rhoades D.F. and Cates R.G. (1976) Toward a general theory of plant antiherbivore chemistry. *Recent Advances in Phytochemistry, 10*, 168–213

Rice E.L. (1984) *Allelopathy*, 2nd Edition, Academic Press, Orlando, FL

Robinson T. (1983) *The organic constituents of higher plants*, 5th Edition, Cordus Press, North Amherst, Mass.

Rogers J.A. and King J. (1972) The distribution and abundance of grassland species in relation to soil aeration and base status. *Journal of Ecology, 60*, 1–18

Romheld V. and Marschner H. (1979) Fine regulation of iron uptake by the Fe-efficient plant *Helianthus annuus*. In *The soil–root interface*, pp. 405–18, Eds J.L. Harley and R. Scott Russell, Academic Press, London

Rorison I.H. (1960) Some aspects of the calcicole–calcifuge problem. I. The effects of competition and mineral nutrition upon seedling growth in the field. *Journal of Ecology, 48*, 585–679

Rorison I.H. (1969) Ecological inferences from laboratory experiments on mineral nutrition. In *Ecological aspects of the mineral nutrition of plants*, pp. 155–76, Eds I.H. Rorison *et al.*, Blackwell, Oxford

Rorison I.H. and Robinson D. (1984) Calcium as an environmental variable. *Plant, Cell and Environment, 7*, 381–90

Rozema J., Luppes E. and Broekman R. (1985) Differential response of salt marsh species to variation of iron and manganese. *Vegetatio, 62*, 293–301

Russell E.W. (1973) *Soil conditions and plant growth*, 10th Edition, Longman, London

Russell R.S. (1977) *Plant root systems*, McGraw-Hill, London

Ruxton B.P. (1969) Rates of weathering of Quaternary volcanic ash in north-east Papua. *Transactions of the IXth International Congress of Soil Science, 4*, 367–76

Saebo S. (1969) On the mechanism behind the effect of freezing and thawing on dissolved phosphorus in *Sphagnum fuscus* peat. *Scientific Reports of the Agricultural College of Norway, 48*, 1–10

Salisbury E.J. (1925) Note on the edaphic succession in some dune soils with special reference to the time factor. *Journal of Ecology, 13*, 322–8

Salisbury F.B. and Ross C.W. (1985) *Plant physiology*, 3rd Edition, Wadsworth Publishing Company, Belmont, CA

Sanchez, P.A. (1976) *Properties and management of soil in the tropics*, John Wiley, New York

Saunders F.E., Mosse B. and Tinker P.B. (Eds) (1975) *Endomycorrhizas*, Academic Press, London

Scholander P.F., Hammel H.T., Bradstreet E.D. and Hemmingsen E.A. (1965) Sap pressure in vascular plants. *Science, 148*, 339–46

Schultz A.M. (1964) The nutrient recovery hypothesis for arctic-microtine cycles. II. Ecosystem variables in relation to arctic-microtine cycles. In *Grazing in terrestrial and marine environments*, pp. 57–68, Ed. D.J. Crisp, Blackwell, Oxford

Sheehy-Skeffington M.J. (1983) *An ecophysiological study of nitrogen budgets in an eastern Irish salt marsh*, PhD Thesis, University of Dublin, Trinity College

Singer C.E. and Havill D.C. (1985) Manganese as an ecological factor in salt marshes. *Vegetatio, 62*, 287–92

Slack A. (1979) *Carnivorous plants*, Ebury Press, London

Smirnoff N. and Stewart G.R. (1985) Stress metabolites and their role in coastal plants. *Vegetatio, 62*, 273–8

Smith S.S.E. (1980) Mycorrhizas of autotrophic higher plants. *Biological Reviews, 55*, 475–510

Sparling J.H. (1967) The occurrence of *Schoenus nigricans* L. in blanket bogs II. Experiments on the growth of *S. nigricans* under controlled conditions. *Journal of Ecology, 55*, 15–31

Specht, R.L. (1963) Dark Island Heath (Ninety Mile Plain, South Australia). VII. The effect of fertilisers on composition and growth; 1950–1960. *Australian Journal of Botany, 11*, 67–94

Specht, R.L. (1969) A comparison of the sclerophyllous vegetation characteristic of Mediterranean-type climates in France, California and Southern Australia. I. Structure, morphology and succession. *Australian Journal of Botany, 17*, 277–92

Specht R.L. (1981) Heathlands. In *Australian vegetation*, pp. 253–75, Ed. R.H. Groves, Cambridge University Press, Cambridge

Specht R.L. and Jones R. (1971) A comparison of the water use by health vegetation at Frankston, Victoria and Dark Island Soak, South Australia.

Australian Journal of Botany, 19, 311–26

Stevens M.E. (1963) Podzol development on a moraine near Juneau, Alaska. *Proceedings of the Soil Science Society of America, 27*, 357–8

Stevens P.R. and Walker T.W. (1970) The chronosequence concept and soil formation. *Quarterly Review of Biology, 45*, 333–50

Stewart G.R. and Ahmad I. (1983) Adaptation to salinity in angiosperm halophytes. In *Metals and micronutrients, uptake and utilisation by plants*, pp. 33–50, Eds D.A. Robb and N.S. Pierpoint, Academic Press, London

Stewart G.R., Lee J.A. and Orebamjo T.O. (1973) Nitrogen metabolism of halophytes. II. Nitrate availability and utilization. *New Phytologist, 72*, 539–46

Stewart G.R. and Orebamjo T.O. (1983) Studies of nitrate utilization by the dominant species of regrowth vegetation of tropical West Africa: a Nigerian example. In *Nitrogen as an ecological factor*, Eds J.A. Lee, S. McNeill and I.H. Rorison, Blackwell, Oxford

Stribley D.P. and Read D.J. (1975) Some nutritional aspects of the biology of ericaceous mycorrhizas. In *Endomycorrhizas*, pp. 196–207, Eds F.E. Saunders, B. Mosse and R.B. Tinker, Academic Press, London

Takhtajan A. (1969) *Flowering plants. Origin and dispersal* (transl. C. Jeffrey), Smithsonian Institute Press, Washington, DC

Talibudeen O. (1981) Precipitation. In *The chemistry of soil processes*, pp. 81–114, Eds D.J. Greenland and M.H.B. Hayes, John Wiley, Chichester

Tate K.R. (1984) The biological transformation of phosphate in soil. *Plant and Soil, 76*, 245–56

Tezuka Y. (1961) Development of vegetation in relation to soil formation in the volcanic island of Oshima, Izu, Japan. *Japanese Journal of Botany, 17*, 371–402

Theodorou C. (1971) The phytase activity of the mycorrhizal fungus *Rhizopogon luteolus. Soil Biology and Biochemistry, 3*, 89–90

Thompson L.M. (1975) *Soils and soil fertility*. 2nd Edition, McGraw-Hill, New York

Tieszen L.L. (Ed.) (1978) *Vegetation and production ecology of an Alaskan Arctic tundra*, Springer, New York

Tinker P.B. (1975) Soil chemistry of phosphorus and mycorrhizal effects on plant growth. In *Endomycorrhizas*, pp. 353–71, Eds F.E. Saunders, B. Mosse and P.B. Tinker, Academic Press, London

Titlyanova A.A. and Bazilevich N.I. (1979) Ecosystem synthesis of meadows — nutrient cycling. In *Grassland ecosystems of the world*, pp. 170–80, Ed. R.T. Coupland, Cambridge University Press, Cambridge

Tubridy M.C. (1983) *A microbiological approach to acid mine waste revegetation*, PhD Thesis, University of Dublin, Trinity College

Tukendorf A. and Baszynski T. (1985) Partial purification and characterisation of copper-binding protein from roots of *Avena sativa* grown on excess copper. *Journal of Plant Physiology, 120*, 57–63

Tuomi J., Niemela P., Houkioja E., Siren S. and Neuvonen S. (1984) Nutrient stress: an explanation for plant anti-herbivore responses to defoliation. *Oecologia, 61* 208–10

Tyler G. (1971) Studies in the ecology of Baltic sea-shore meadows. IV.

Distribution and turnover of organic matter and minerals in a shore meadow ecosystem. *Oikos, 22*, 265–91

Ure A.M. and Barrow M.L. (1982) The elemental constituents of soils. In *Environmental chemistry*, Vol. 2, Ed. H.J.M. Bowen, Royal Society of Chemistry, London

Valiela I. and Teal J.M. (1979) The nitrogen budget of a salt marsh ecosystem. *Nature (London), 280*, 652–6

Vallee B.L. and Ulmer D.U. (1972) Biochemical effects of mercury, cadmium and lead. *Annual Review of Biochemistry, 41*, 91–128

Van Breemen N. (1973) Soil forming process in acid sulphate soils. In *Acid sulphate soils*, pp. 66–131, Ed. H. Dorst, Institute for Land Reclamation and Improvement, Wageningen

Warcup J.H. (1975) A culturable *Endogone* associated with eucalypts. In *Endomycorrhizas*, pp. 53–63, Eds F.E. Saunders, B. Mosse and P.B. Tinker, Academic Press, London

Watanabe I. and Brotonegoro S. (1981) Paddy fields. In *Nitrogen fixation, Vol. 1, Ecology*, pp. 242–63, Ed. W.J. Broughton, Oxford University Press, Oxford

Webber P.J. (1980) Spatial and temporal variation of the vegetation and its productivity. In *Vegetation and production ecology of an Alaskan Arctic tundra*, pp. 37–112, Ed. L.L. Tieszen, Springer, New York

Wein W. and McLean D.A. (1983) *The role of fire in northern circumpolar ecosystems*, Scope 18, Wiley, Chichester

Whittaker R.H. (1960) Vegetation of the Siskiyou Mountains, Oregon and California. *Ecological Monographs, 30*, 279–338

Whittaker R.H. (1975) *Communities and ecosystems*, 2nd Edition, Macmillan Publishing Co., New York

Wilkins D.A. (1957) A technique for the measurement of lead tolerance in plants. *Nature (London), 180*, 37–8

Williams R.F. (1955) Redistribution of mineral elements during development. *Annual Review of Plant Physiology, 6*, 25–42

Williams S.T. and Mayfield C.I. (1971) Studies in the ecology of actinomycetes in soil. III. The behaviour of neutrophilic streptomycetes in acid soil. *Soil Biology and Biochemistry, 3*, 107–208

Willis A.J. (1963) Braunton Burrows: the effects on the vegetation of the addition of mineral nutrients to the dune soils. *Journal of Ecology, 51*, 353–74

Woodell S.R.J., Mooney H.A. and Hill A.J. (1969) The response of *Larrea divaricata* (creosote bush) to rainfall in California. *Journal of Ecology, 57*, 37–44

Woolhouse H.W. (1969) The acid phosphatases of plant roots. In *Ecological aspects of the mineral nutrition of plants*, pp. 357–80, Eds I.H. Rorison *et al.*, Blackwell, Oxford

Woolhouse H.W. (1983) Toxicity and tolerance in the responses of plants to metals. In *Physiological plant ecology III*, pp. 245–300, Eds O.L. Lange, P.S. Nobel, C.B. Osmond and H. Ziegler, Springer, Berlin

Yakuwa R. (1945) Uber die Bodentemperaturen in dem verschiedenen Bodenarten in Hokkaido. *Geophysical Magazine (Tokyo), 14*, 1–12

Young R.S. (1979) *Cobalt in biology and biochemistry*, Academic Press, London

Index